New Horizon of Plasmonics and Metamaterials

New Horizon of Plasmonics and Metamaterials

Special Issue Editors

Masafumi Kimata
Shinpei Ogawa

MDPI • Basel • Beijing • Wuhan • Barcelona • Belgrade

Special Issue Editors
Masafumi Kimata
Ritsumeikan University
Japan

Shinpei Ogawa
Mitsubishi Electric Corporation
Japan

Editorial Office
MDPI
St. Alban-Anlage 66
4052 Basel, Switzerland

This is a reprint of articles from the Special Issue published online in the open access journal *Materials* (ISSN 1996-1944) from 2018 to 2020 (available at: https://www.mdpi.com/journal/materials/special_issues/metamaterials_plasmonics).

For citation purposes, cite each article independently as indicated on the article page online and as indicated below:

LastName, A.A.; LastName, B.B.; LastName, C.C. Article Title. *Journal Name* **Year**, *Article Number*, Page Range.

ISBN 978-3-03936-014-7 (Hbk)
ISBN 978-3-03936-015-4 (PDF)

Contents

About the Special Issue Editors

Masafumi Kimata (Prof.) received his MS degree from Nagoya University in 1976, and received his Ph.D. degree from Osaka University in 1992. He joined Mitsubishi Electric Corporation in 1976, and retired from Mitsubishi Electric in 2004. Currently, he is a professor at Ritsumeikan University, where he continues his research on MEMS-based uncooled infrared focal plane arrays and type-II superlattice infrared focal plane arrays. He is a fellow of SPIE.

Shinpei Ogawa (Dr.) received his BE and ME degrees and Ph.D. from the Department of Electronic Science and Engineering, Kyoto University, Japan, in 2000, 2002, and 2005, respectively. He joined Mitsubishi Electric Corporation in 2005. He works on various devices including RF-MEMS devices, advanced infrared sensors using plasmonics and metamaterials, and graphene, or other 2D materials-based photodetectors. He is currently a unit leader of plasmonics, metamaterials, and graphene research for novel optoelectronic devices.

materials

MDPI

Editorial

Special Issue: New Horizon of Plasmonics and Metamaterials

Shinpei Ogawa [1,*] and Masafumi Kimata [2]

1 Advanced Technology R&D Center, Mitsubishi Electric Corporation, 8-1-1 Tsukaguchi-Honmachi, Amagasaki, Hyogo 661-8661, Japan

2 College of Science and Engineering, Ritsumeikan University, 1-1-1 Noji-higashi, Kusatsu, Shiga 525-8577, Japan; kimata@se.ritsumei.ac.jp

* Correspondence: Ogawa.Shimpei@eb.MitsubishiElectric.co.jp; Tel.: +81-6-6497-6436

Received: 2 April 2020; Accepted: 8 April 2020; Published: 9 April 2020

Abstract: Plasmonics and metamaterials are growing fields that consistently produce new technologies for controlling electromagnetic waves. Many important advances in both fundamental knowledge and practical applications have been achieved in conjunction with a wide range of materials, structures and wavelengths, from the ultraviolet to the microwave regions of the spectrum. In addition to this remarkable progress across many different fields, much of this research shares many of the same underlying principles, and so significant synergy is expected. This Special Issue introduces the recent advances in plasmonics and metamaterials and discusses various applications, while addressing a wide range of topics in order to explore the new horizons emerging for such research.

Keywords: plasmonics; metamaterials; metasurfaces; polarization control; infrared sensors

---

Plasmonics and metamaterials are both fields of study that have received increasing attention and that constantly produce new means of modifying electromagnetic waves. Surface plasmon polaritons (SPPs) are electromagnetic waves at the interface between a metal and a dielectric that are excited by the coupling of photons and electrons. In general, SPPs having wavelengths from the ultraviolet to the far infrared (IR) regions of the spectrum can be produced. Metamaterials are engineered artificial structures exhibiting unconventional physical properties that cannot be achieved using standard materials. Recently, there has been significant interest in metasurfaces based on two dimensional metamaterials because of the potential of such surfaces to manipulate photons. These two technologies can be combined, typically in conjunction with optical wavelengths, to produce unique properties. Consequently, many results that are important both in terms of our fundamental understanding of these phenomena and in terms of actual applications have been obtained, using a wide range of materials, structures and wavelengths spanning the ultraviolet to the microwave. Interestingly, despite the numerous fields in which these technologies have been investigated, much of this research has many common principles and so could lead to progress via synergistic effects and collaborations.

This Special Issue, "New Horizon of Plasmonics and Metamaterials", brings together eight articles and one review that capture and summarize the recent activity and developments in this field as well as practical applications over a wide range of topics, so as to explore the new possibilities emerging for these fields.

To date, three studies have demonstrated polarization control using metasurfaces or metagratings in the GHz frequency range. Shi et al. [1] employed metasurfaces with multiple layers in association with the generation of vortex beams and conversion via cross-polarization. Shi's work involved the development of dual metasurfaces for the purpose of polarization conversion as a means of producing beams carrying orbital angular momenta with four different orders (such that $l = +1, l = +2, l = -1$ and $l = -2$). The data from this work assisted in the realization of polarimetric and super resolution imaging. Song et al. [2] reported a means of performing the multifunctional manipulation of polarization, based on using a dielectric grating containing periodic arrangements of meta-atoms to produce metagratings. These devices acted as highly efficient waveplates with dual modes and multiple functions, and were able to exhibit a number of different functions. These functions included circular or linear cross-polarization and linear-to-circular polarization conversions as well as mirroring with chirality preservation for a variety of frequency bands, together with significant angular invariance. Such properties could potentially be useful in conjunction with additional ranges of frequencies so as to fabricate small optical polarization control devices for optical, radar and telecommunications applications. Li et al. [3] developed a cross-polarization converter capable of dual-band operation with a transmissive unit cell design with multiple layers, based on a transmit array with aperture coupling. In this device, co-polarized transmittance is greatly reduced (to less than 0.005), giving a ratio of polarization conversion that is almost ideal. This converter has potential applications in telecommunications, radar systems and antennae.

Two studies have examined plasmonic effects in waveguides in conjunction with wavelengths in the visible and near-IR spectral range. Zhang et al. [4] reported a metal-dielectric-metal plasmonic device with gain-assisted operation that exhibited improved slow light operation as a result of a transparency effect related to plasmons. In this system, the optical delay and transmission of slow light can both be improved by adjusting the gain power. In addition, incorporating an additional gain disk cavity allows for the enhanced introduction of slow light via a double-channel. This device could have uses in optical switching, nanolasers and biosensing. Wang et al. [5] produced a metal-insulator-metal (MIM) system incorporating coupled hetero-cavities that generates and tunes three Fano resonances. The multiple Fano resonances are obtained via separate mechanisms involving cavity–cavity coupling and can be considered to represent two different types of resonance. Each type can be tuned separately by modifying various parameters related to the cavities, so as to allow tunable modulation of the Fano resonances. This technology has potential applications in slow-light devices, filters, nanosensors and modulators.

Other work has examined the use of this technology operating at wavelengths at visible and IR wavelengths. Kanamori et al. [6] developed a miniature spectroscope incorporating 25 color sensors in association with Si photodiodes and color filters made of metamaterials. These filters comprised metal gratings exhibiting guided mode resonance, with subwavelength periodic two-dimensional morphologies. The spectral sensitivity of this device was determined to exhibit a peak wavelength that had a linear correlation with the period of the grating. Upon irradiation with monochromatic light at various wavelengths, the incident light's spectral characteristics could be recovered from the signals generated by the color sensors. This metamaterial filter technology could be applied to the fabrication of image sensors operating in multiple colors. Ogawa et al. [7] researched uncooled IR sensors operating in selective polarization and wavelength capacities to design a means of removing undesirable modes while employing a number of plasmonic metamaterial absorbers (PMAs) and applying a reference pixel and a subtraction process. This same approach has possible applications in a number of different uncooled IR sensors. Ogawa et al. [8] also published a review of the different MIM-PMAs that have been reported. This review discusses the history of these devices together with their basic physical principles and modes of operating, while also addressing the research that will be necessary in the future to elucidate design aspects and allow different applications. The technology discussed by Ogawa could be used in many wavelength regions, including the microwave, terahertz and ultraviolet ranges of the electromagnetic spectrum.

Sabouni-Zawadzka et al. [9] examined a new cellular metamaterial having a simplex tensegrity morphology. Their work involved the use of three different tensegrity lattices with varying geometric structures and assessed six different deformation modes: low and high (double) shear along with soft, stiff and medium extensional. Both unimode and close to bimode lattices were reported, based on a classification system for extreme advanced materials.

The brief summary above introduces a wide range of topics, including optics, radiofrequency engineering and mechanics. This variety of research illustrates the rapid progress that has occurred in the field of plasmonics and metamaterials. We hope that this Special Issue will inspire researchers to continue to perform groundbreaking research in this area.

Funding: This research received no external funding.

Conflicts of Interest: The authors declare no conflict of interest.

References

1. Shi, H.; Wang, L.; Zhao, M.; Chen, J.; Zhang, A.; Xu, Z. Transparent Metasurface for Generating Microwave Vortex Beams with Cross-Polarization Conversion. *Materials* **2018**, *11*, 2448. [CrossRef] [PubMed]
2. Song, K.; Ji, R.; Shrestha, D.; Ding, C.; Liu, Y.; Zhu, W.; He, W.; Liu, H.; Guo, Y.; Tang, Y.; et al. High-Efficiency and Wide-Angle Versatile Polarization Controller Based on Metagratings. *Materials* **2019**, *12*, 623. [CrossRef] [PubMed]
3. Li, J.; Feng, J.; Li, B.; Shi, H.; Zhang, A.; Chen, J. Dual-Band Transmissive Cross-Polarization Converter with Extremely High Polarization Conversion Ratio Using Transmitarray. *Materials* **2019**, *12*, 1827. [CrossRef] [PubMed]
4. Zhang, Z.; Yang, J.; He, X.; Han, Y.; Zhang, J.; Huang, J.; Chen, D.; Xu, S. Active Enhancement of Slow Light Based on Plasmon-Induced Transparency with Gain Materials. *Materials* **2018**, *11*, 941. [CrossRef] [PubMed]
5. Wang, Q.; Ouyang, Z.; Lin, M.; Liu, Q. Independently Tunable Fano Resonances Based on the Coupled Hetero-Cavities in a Plasmonic MIM System. *Materials* **2018**, *11*, 1675. [CrossRef] [PubMed]
6. Kanamori, Y.; Ema, D.; Hane, K. Miniature Spectroscopes with Two-Dimensional Guided-Mode Resonant Metal Grating Filters Integrated on a Photodiode Array. *Materials* **2018**, *11*, 1924. [CrossRef] [PubMed]
7. Ogawa, S.; Takagawa, Y.; Kimata, M. Elimination of Unwanted Modes in Wavelength-Selective Uncooled Infrared Sensors with Plasmonic Metamaterial Absorbers using a Subtraction Operation. *Materials* **2019**, *12*, 3157. [CrossRef] [PubMed]
8. Ogawa, S.; Kimata, M. Metal-Insulator-Metal-Based Plasmonic Metamaterial Absorbers at Visible and Infrared Wavelengths: A Review. *Materials* **2018**, *11*, 458. [CrossRef] [PubMed]
9. Al Sabouni-Zawadzka, A.; Gilewski, W. Soft and Stiff Simplex Tensegrity Lattices as Extreme Smart Metamaterials. *Materials* **2019**, *12*, 187. [CrossRef] [PubMed]

materials

MDPI

Article

Transparent Metasurface for Generating Microwave Vortex Beams with Cross-Polarization Conversion

Hongyu Shi [1], Luyi Wang [1], Mengran Zhao [1], Juan Chen [2,*], Anxue Zhang [1] and Zhuo Xu [3]

1. School of Electronic and Information Engineering, Xi'an Jiaotong University, Xi'an 710049, China; honyo.shi1987@gmail.com (H.S.); bigcrash@stu.xjtu.edu.cn (L.W.); zmr1993@stu.xjtu.edu.cn (M.Z.); anxuezhang@xjtu.edu.cn (A.Z.)
2. Xi'an Jiaotong University Shenzhen Research School, Shenzhen 518057, China
3. Electronic Materials Research Laboratory, Key Laboratory of the Ministry of Education, Xi'an Jiaotong University, Xi'an 710049, China; xuzhuo@xjtu.edu.cn

* Correspondence: chen.juan.0201@xjtu.edu.cn

Received: 25 October 2018; Accepted: 26 November 2018; Published: 3 December 2018

Abstract: In this paper, metasurfaces with both cross-polarization conversion and vortex beam-generating are proposed. The proposed finite metasurface designs are able to change the polarization of incident electromagnetic (EM) waves to its cross-polarization. In addition, they also can modulate the incidences into beams carrying orbital angular momentum (OAM) with different orders ($l = +1, l = +2, l = -1$ and $l = -2$) by applying corresponding transmission phase distribution schemes on the metasurface aperture. The generated vortex beams are at 5.14 GHz. The transmission loss is lower than 0.5 dB while the co-polarization level is -10 dB compared to the cross-polarization level. The measurement results confirmed the simulation results and verified the properties of the proposed designs.

Keywords: vortex beam; polarization conversion; orbital angular momentum

1. Introduction

The orbital angular momentum of electromagnetic waves has been explored in recent years for its potential applications in wireless communications [1–3] and imaging [4,5]. EM wave carrying orbital angular momentum has a helical wavefront and an amplitude singularity in the propagating direction. The helical wavefront can be expressed by the term $exp(il\Phi)$, where Φ is the azimuthal angle and l is the topological charge. The topological charge corresponds to the OAM mode and, theoretically, the OAM mode is vast.

The OAM in EM waves is typically generated using techniques like spiral phase plates [6,7], spiral reflectors [8], antennas [9–11], dieletric resonators [12], computer-induced holograms [13], transformation electromegnetics [14] and metasurfaces [15,16]. The common idea in these techniques is to introduce the desired phase distribution on the radiation aperture. The spiral phase plate method gives the incident wave different phase retardation according to the term $exp(il\Phi)$ by modulating the length of the wave path in corresponding areas. The antenna array approach usually use a circular antenna array to excite array elements with the same amplitude but different phases.

Compared to these methods, metasurfaces for generation of beams carrying OAM have advantages including low profile and simple EM wave control, i.e., the magnitude/phase/polarization of the EM waves can be manipulated simply by changing the shape, geometry, size, orientation and arrangement of the structures [17]. Reflective metasurfaces were used to generate single and double mode vortex beams in mircrowave [18–21] and terahertz regimes [22]. An active transparent metasurface was proposed for generating EM beams carrying OAM in the microwave frequency range [23]. However, these designs only focus on the OAM controlling of microwaves leaving the polarization state of the transmitted wave the same as that of the incidence. Simultaneously control the polarization and OAM of EM waves can enhance the performances of OAM beams in applications like radar imaging. Recently, metasurfaces using Geometric-Phase were applied for simultaneous OAM and spin angular momentum control [24–26]. These techniques impart a new degree of freedom to EM wave control and pave a way for future applications.

In this paper, multi-layered metasurfaces generating OAM beams with efficient linear polarization conversion were proposed. The patches on the outer sides of the designed metasurface receive and re-radiate the incident wave, respectively. The cross-polarized transmission is higher than −1 dB around 5.15 GHz with an extremely low co-polarized transmission below −35 dB. The transmission phase can be fully controlled by the length of the stripline in the middle layer. By arranging the metasurface unit cells according to desired phase distributions, the proposed metasurfaces can generate EM beams carrying different modes of OAM. This design method was demonstrated by both simulation and measurement.

This paper is organized as follows: Section 2 presents the design of the unit cells and the metasurfaces. In this section, detailed geometries of the unit cell are introduced, the characteristics of the unit cell are shown and the metasurface designs are presented. Section 3 presents the simulated and measured results of the metasurfaces, which verify the designs and show OAM generation with polarization conversion. In Section 4, the conclusions are drawn and the originality of this work is presented.

2. Design of The Metasurface

A flowchart illustrating the main goal and the adopted methodology of this study is presented in Figure 1. The unit cell pattern of the proposed multi-layered laminated metasurface is depicted in Figure 2, where the gray parts represent the substrate Rogers 4350 B with ϵ_r = 3.48 and tanδ = 0.0037. The yellow parts represent the metal structures with a thickness of 0.035 mm. The top and bottom layers of the metasurface are circular patches which can couple or decouple the incident wave with a cross-polarization conversion. The middle layer of the metasurface is a stripline structure and is separated from the top and bottom layers by the first and second ground layers, respectively. Two vias connect the two ends of the stripline to the top and bottom layers, respectively. The geometric dimensions are p = 17.92 mm, R = 16.64 mm, r = 0.8 mm, w = 1.2 mm, $f_x = f_y$ = 3 mm, h_1 = 1.524 mm and h_2 = 0.254 mm. The length of the stripline S varies from 1.66 mm to 8.49 mm to achieve a 360° transmission phase control.

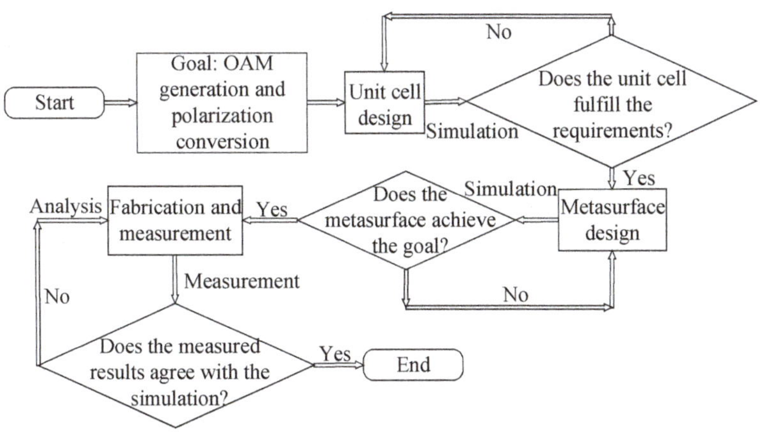

Figure 1. Flowchart illustrating the main goal and the adopted methodology of this paper.

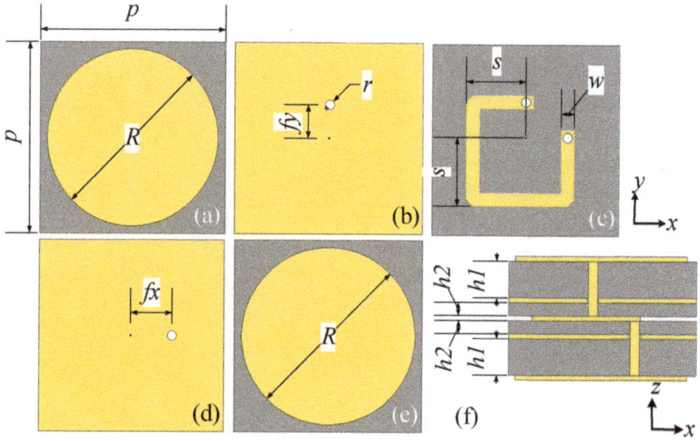

Figure 2. Geometry of the unit cell: (**a**) Top layer. (**b**) First ground layer. (**c**) Middle layer. (**d**) Second ground layer. (**e**) Bottom layer. (**f**) Side view.

The unit cell design takes its inspiration from the patch antenna. The top/bottom layer couples the y-polarized/x-polarized incident wave into the guided mode in the stripline structure and then, the bottom/top layer decouples the guided wave into the x-polarized/y-polarized free space propagation. It is by selecting the positions of the vias in orthogonal direction (e.g., in x and y directions), that the polarization of the transmitted wave is converted.

The simulated distributions of the electric field component perpendicular to the unit cell (i.e., E_z) on top and bottom layers are shown in Figure 3a,b, respectively. The incidences excite a y-polarized dipolar mode on the top layer, where the guided mode travels through the stripline structure to the bottom layer and excite a x-polarized dipolar mode, therefore converting the polarization of the transmitted waves. In addition, the guided mode experiences different phase delay when the length of the stripline varies. Also, the energy loss in the stripline structure is small and consistent regardless of its lengths. Therefore, the transmission phase can be controlled by the length of the stripline, which allows different phase distributions for different OAM beam-generating, while the transmission loss is small and stable. Notably, due to only the top and bottom layers resonate, this design has potentials to obtain low insert loss.

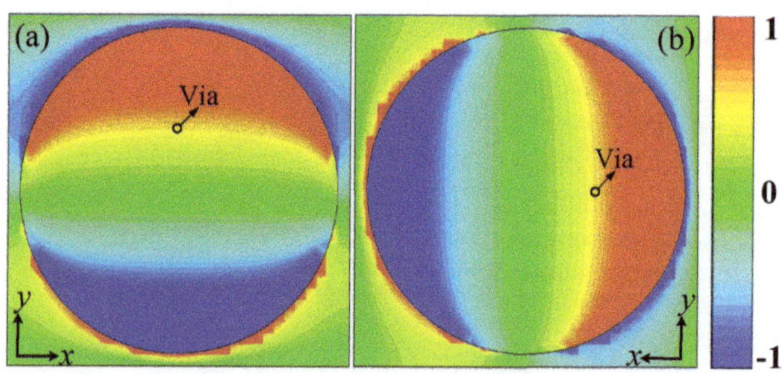

Figure 3. The simulated distributions of the electric field component perpendicular to the unit cell (i.e., E_z): (**a**) Top layer. (**b**) Bottom layer.

The unit cell models were simulated by the commercial software CST Microwave Studio (Version 2016, Computer Simulation Technology GmbH, Darmstadt, Germany) using periodic boundary in x and y directions. The simulated transmission amplitudes and phases are shown in Figure 4a,b, respectively. The transmission phase data at 4.8–5.5 GHz are given because, at other frequency ranges, the curves are confused and not of main concern in this paper. For a y-polarized incidence propagating along $-z$ direction, the transmitted wave is x-polarized and the cross-polarized transmission amplitudes with different S are higher than -0.5 dB at 5.14 GHz, and from 4.9 GHz to 5.4 GHz, the transmittances are higher than -3 dB which indicates a 50% power efficiency, as shown in Figure 4a. The co-polarized transmittances are below -35 dB from 4.9 GHz to 5.4 GHz, indicating an extremely high polarization conversion efficiency, compared with [27,28]. At 5.14 GHz, the co-polarized transmittance is below -38.2 dB. Eight stripline lengths were selected with a transmission phase step of 45° to cover a 360° phase difference, as shown in Figure 4b. The selected stripline lengths are 1.656 mm, 2.62 mm, 3.62 mm, 4.58 mm, 5.57 mm, 6.54 mm, 7.53 mm and 8.49 mm.

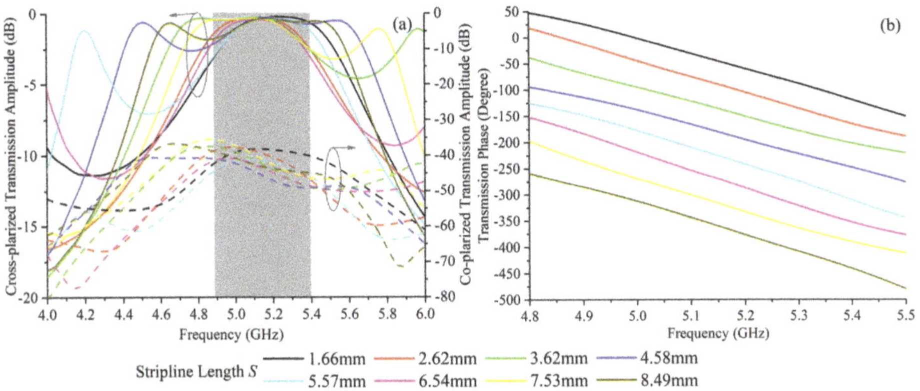

Figure 4. The simulated transmittance of the unit cell with different stripline lengths S (as in Figure 2c): (**a**) Amplitude. (**b**) Phase.

The helical wavefront of vortex beams can be expressed by the term $exp(il\Phi)$, where Φ is the azimuthal angle and l is the topological charge. Therefore, EM beams carrying OAM with an order of l experiences an azimuthal phase change of $|\ l\ | \times 360°$. The sign of the OAM order l defines the helicity of the vortex beam phase distribution. To generate vortex beam-carrying OAM of orders ±1 and ±2, two transmission phase distributions at 5.14 GHz with phase steps of 45° and 90°, respectively were designed and shown in Figure 5a,b, which depict the desired transmission phase with regard to different positions on metasurfaces. For wave propagating along $-z$ and z directions, these two designs have opposite helicities and generate EM beams carrying OAM with orders of 1/2 and $-1/-2$, respectively.

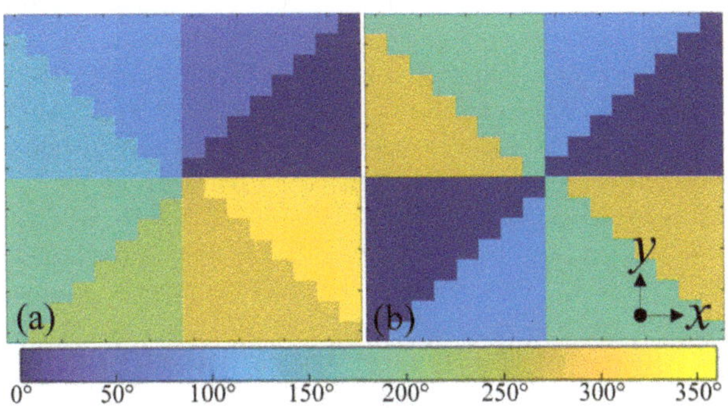

Figure 5. The front view of the transmission phase distribution schemes at 5.14 GHz for generating beams carrying OAM of different orders: (**a**) $l = +1$. (**b**) $l = +2$.

The two finite full structure models containing 16 × 16 unit cells are shown in Figure 6. The discretization of the metsurface is done according to the transmission phase distributions in Figure 5. The target frequency of the metasurfaces is at 5.14 GHz. It is worth pointing out that the potential applications for radar imaging can be in X/C/S band and the metasurface design can be easily tuned to other frequencies as well. The models were built up and simulated by CST Microwave studio with a Gaussian beam as excitation with a minimum beam radius of 100 mm on the metasurface. The average simulation time in a server with 256 GB memory and Intel Xeon E5 CPU is about 6 h. About 30 GB memory is used. Gaussian beam, compared with plane wave, reduces the slight amount of diffractions of the EM waves at the margins, while the phase profile of the transmitted beams are the same. Also, the margins of the metasurfaces have metal sheet in ground layers to further avoid diffractions. The used unit cells for realizing the desired transmission phase distribution designs in Figure 5a,b with phase steps of 45° and 90° respectively are selected from Figure 4b. Eight kinds of unit cell with stripline lengths of 1.656 mm, 2.62 mm, 3.62 mm, 4.58 mm, 5.57 mm, 6.54 mm, 7.53 mm and 8.49 mm are selected for Figure 5a while four kinds of unit cell with stripline lengths of 1.656 mm, 3.62 mm, 5.57 mm and 7.53 mm are selected for Figure 5b.

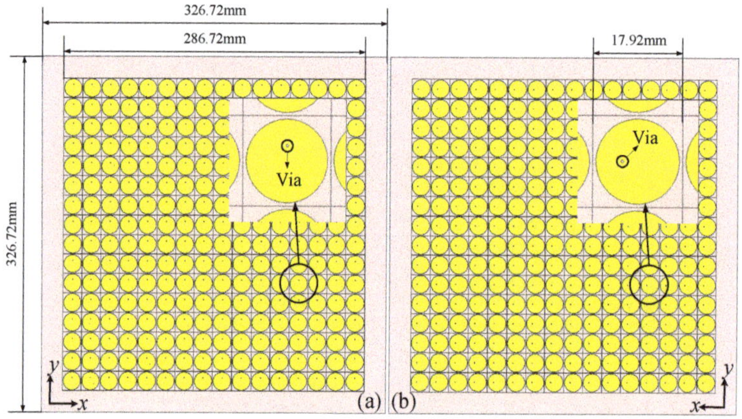

Figure 6. The simulation model of the proposed metasurface: (**a**) Front view. (**b**) Back view.

3. Results

The simulated transmitted electric field distributions at a transverse plane 250 mm away from the metasurface are depicted in Figure 7. Figure 7a–d,e–h show the transmitted electric field distributions with incidences propagating along $-z$ direction with a y-polarization and along z direction with a x-polarization, respectively. Figure 7a,c,e,g show the normalized transmitted electric field amplitude distributions at 5.14 GHz with design schemes in Figure 5a,b, respectively. Amplitude nulls can be observed due to the phase singularity at the center of OAM carrying beams, and along with the donut-shaped field distribution verified the characteristic of the vortex beams.

The transmitted phase distributions are shown in Figure 7b,d,f,h. The phase accumulations along a full circular path around the beam null in Figure 7b,d are 2π and -2π, which indicates OAM orders of $+1$ and -1, respectively. Figure 7f,h depict 4π and -4π phase accumulations along a full circular path and therefore indicate OAM orders of $+2$ and -2. Thus, by using the proposed structure, the designed metasurfaces can simultaneously convert the polarization of the incident wave and generate vortex beams carrying OAM of four different orders, which has great potentials for radar imaging applications.

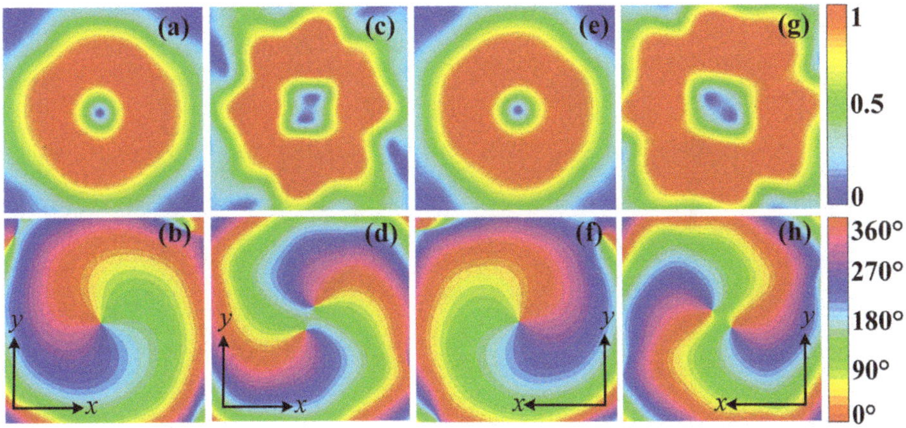

Figure 7. Simulated cross-polarized electric field distributions of the transmitted OAM carrying beams at a transverse plane 250 mm away from the metasurface: (**a**) Amplitude and (**b**) phase distributions for OAM order of $l = +1$. (**c**) Amplitude and (**d**) phase distributions for OAM order of $l = +2$. (**e**) Amplitude and (**f**) phase distributions for OAM order of $l = -1$. (**g**) Amplitude and (**h**) phase distributions for OAM order of $l = -2$.

The proposed metasurface was fabricated using PCB processing as shown in Figure 8. The overall size of the fabricated sample is 326.72 mm × 326.72 mm with a thickness of 4.35 mm. Vias connecting the middle layer to the top and bottom layers are fabricated by back drilling leaving two holes on the top and bottom layers of each unit cell. The back drill holes and prepregs have been considered in the simulations and have little influences on the metasurfaces performance.

The fabricated metasurfaces were measured using a vector network analyzer Agilent E8363b (Keysight Technologies, California, United States) and a two-dimensional near field scanning measurement system. The metasurface was placed between a lens horn antenna (used as the excitation) and a WR-229 open-ended rectangular waveguide probe (used for receiving the OAM carrying beams). The measurement was conducted in the anechoic chamber. The response calibration was used to eliminate the effect of external noise during the measurement. The polarization conversion was confirmed by receiving and analyzing the co-polarization and cross-polarization components of the EM waves, which was realized by rotating the open-ended rectangular waveguide probe. A schema of the measurement devices and settings is depicted in Figure 9.

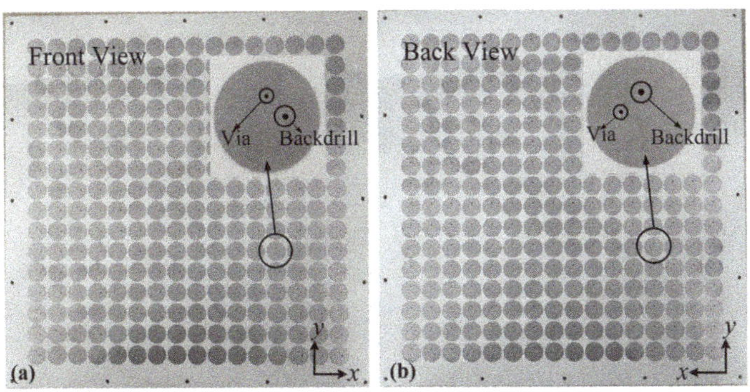

Figure 8. Photos of the fabricated metasurface: (**a**) Front view. (**b**) Back view.

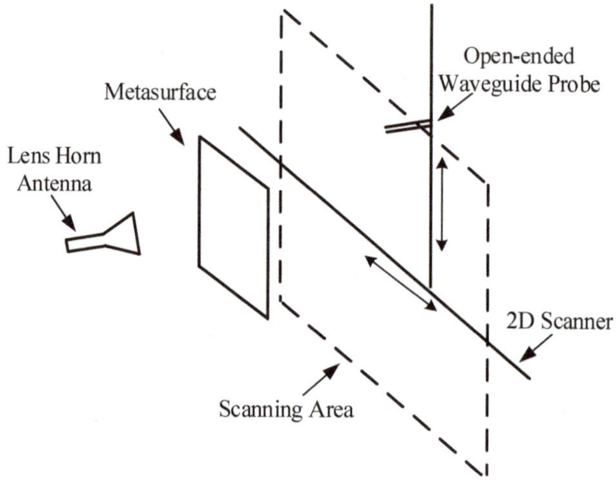

Figure 9. Schema depicting the measurement devices and settings.

The measured amplitude and phase distributions of the cross-polarized transmitted electric field at a transverse plane 250 mm away from the metasurface are shown in Figure 10. Figure 10a–d show the amplitudes and phases of the x-polarized transmitted wave with a y-polarized excitation propagating along $-z$ direction. Figure 10a,c depict an amplitude null at the field center. Figure 10b,d show phase accumulations of 2π and 4π along a full circular path, indicating OAM orders of $+1$ and $+2$, respectively. Figure 10e–h show the amplitudes and phases of the y-polarized transmitted wave with an x-polarized excitation propagating along z direction. The amplitude distributions shown in Figure 10e,g show amplitude nulls level at 0.01 compared to the maximum value. Due to the deviations in fabrication and measurement, the perfect offset of amplitude at the center may be compromised. Still, -20 dB nulls level is satisfying [29]. The phase distributions in Figure 10f,h show -2π and -4π phase accumulations along a full circular path, indicating OAM order of -1 and -2, respectively.

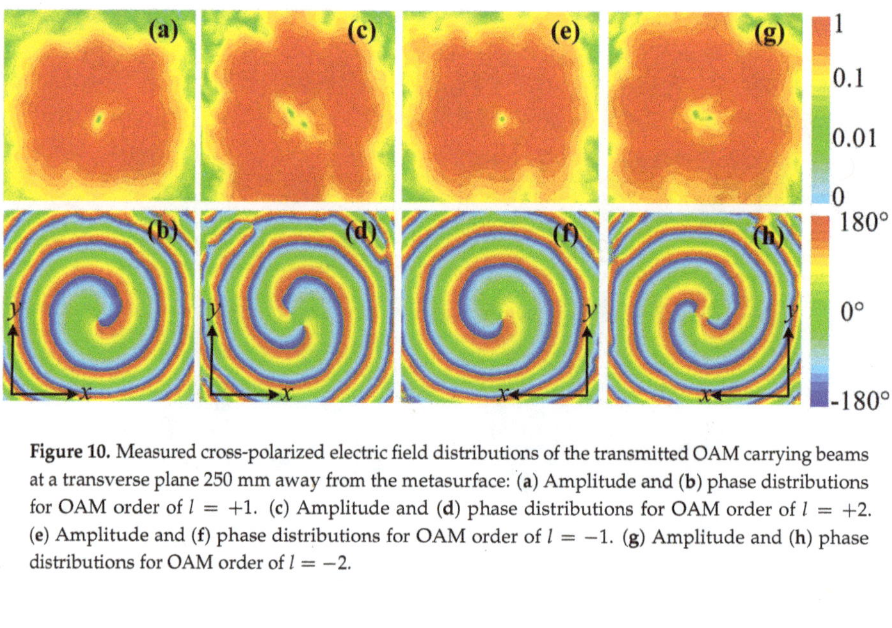

Figure 10. Measured cross-polarized electric field distributions of the transmitted OAM carrying beams at a transverse plane 250 mm away from the metasurface: (**a**) Amplitude and (**b**) phase distributions for OAM order of $l = +1$. (**c**) Amplitude and (**d**) phase distributions for OAM order of $l = +2$. (**e**) Amplitude and (**f**) phase distributions for OAM order of $l = -1$. (**g**) Amplitude and (**h**) phase distributions for OAM order of $l = -2$.

The simulated and measured amplitudes of the co-polarized transmitted wave are shown in Figure 11a,b, respectively. For each condition, the co-polarized transmissions are randomly distributed with a low amplitude. In the simulation results, the co-polarized electric fields amplitude level is lower than 0.06, while the measured results show co-polarization level lower than 0.15. The discrepancy between simulation and measurement comes from fabrication deviations and the slight amount of diffracted EM waves. However, compared with the cross-polarization level, the co-polarization level is low and does not affect the generated cross-polarized vortex beams. The co-polarization level can be enhanced if absorbers are placed around the metasurface.

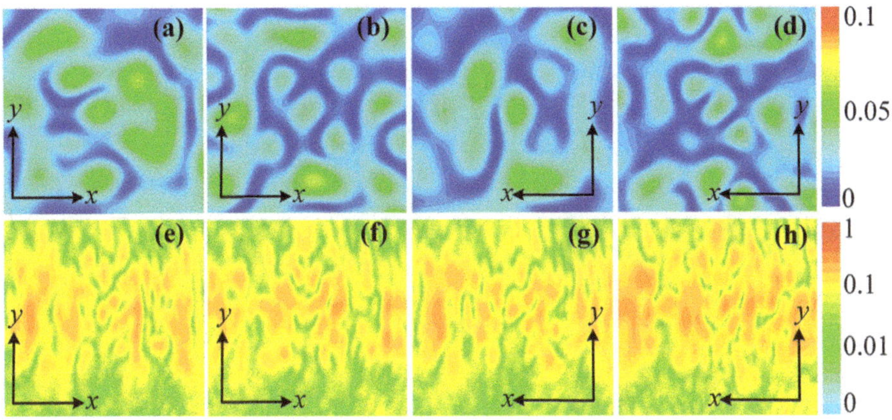

Figure 11. Simulated co-polarized electric fields amplitude distributions for different OAM orders: (**a**) $l = +1$. (**b**) $l = +2$. (**c**) $l = -1$. (**d**) $l = -2$. Measured co-polarized electric fields amplitude distributions for different OAM orders: (**e**) $l = +1$. (**f**) $l = +2$. (**g**) $l = -1$. (**h**) $l = -2$.

4. Conclusions

In conclusion, two polarization conversion metasurfaces generating four different orders of OAM carrying beams ($l = +1, l = +2, l = -1$ and $l = -2$) were designed and fabricated. The simulation and measurement results are in good agreement with each other. The multi-layered unit cells we proposed realize full phase control, low transmission loss, high polarization conversion efficiency and can be easily tuned to any frequencies of interest. By manipulating the transmission phase distributions on the metasurface aperture, the transmitted beams can carry OAM of four different orders, which has potentials for super resolution imaging. In addition, the polarizations of the transmitted waves were efficiently converted, which may further enhance the performances in applications, for example, imaging polarization dependent objects.

Author Contributions: Conceptualization, H.S.; methodology, H.S. and L.W.; validation, H.S.; formal analysis, H.S. and L.W.; investigation, L.W. and M.Z.; resources, A.Z.; data curation, L.W.; writing—review and editing, H.S.; visualization, H.S. and L.W.; supervision, H.S. and J.C.; project administration, J.C., A.Z. and Z.X.; funding acquisition, H.S.

Funding: This research was funded by National Natural Science Foundation of China grant number 61871315, in part by Technology Program of Shenzhen grant number JCYJ20170816100722642, in part by the Natural Science Foundation of Guangdong Province, China grant number 2018A030313429 and in part by the China Postdoctoral Science Foundation under Grant 2015M580849. The APC was funded by 61871315.

Conflicts of Interest: The authors declare no conflict of interest.

References

1. Yan, Y.; Xie, G.D.; Lavery, M.P.J.; Huang, H.; Ahmed, N.; Bao, C.J.; Ren, Y.X.; Cao, Y.W.; Li, L.; Zhao, Z.; et al. High-capacity millimetre-wave communications with orbital angular momentum multiplexing. *Nat. Commun.* **2014**, *5*, 4876. [CrossRef] [PubMed]
2. Hui, X.N.; Zheng, S.L.; Chen, Y.L.; Hu, Y.P.; Jin, X.F.; Chi, H.; Zhang, X.M. Multiplexed Millimeter Wave Communication with Dual Orbital Angular Momentum (OAM) Mode Antennas. *Sci. Rep.* **2015**, *5*, 10148. [CrossRef] [PubMed]
3. Mahmouli, F.E.; Walker, S.D. 4-Gbps Uncompressed Video Transmission over a 60-GHz Orbital Angular Momentum Wireless Channel. *IEEE Wirel. Commun. Lett.* **2013**, *2*, 223–226. [CrossRef]
4. Liu, K.; Cheng, Y.Q.; Yang, Z.C.; Wang, H.Q.; Qin, Y.L.; Li, X. Orbital-Angular-Momentum-Based Electromagnetic Vortex Imaging. *IEEE Antennas Wirel. Propag. Lett.* **2015**, *14*, 711–714. [CrossRef]
5. Liu, K.; Cheng, Y.Q.; Gao, Y.; Li, X.; Qin, Y.L.; Wang, H.Q. Super-resolution Radar Imaging Based on Experimental OAM Beams. *Appl. Phys. Lett.* **2017**, *116*, 164102. [CrossRef]
6. Cheng, L.; Hong, W.; Hao, Z.C. Generation of Electromagnetic Waves with Arbitrary Orbital Angular Momentum Modes. *Sci. Rep.* **2014**, *5*, 4814. [CrossRef]
7. Chen, Y.; Zheng, S.; Li, Y.; Hui, X.; Jin, X.; Chi, H.; Zhang, X. A Flat-Lensed Spiral Phase Plate Based on Phase-shifting Surface for Generation of Millimeter-wave OAM Beam. *IEEE Antennas Wirel. Propag. Lett.* **2015**, *15*, 1156–1158. [CrossRef]
8. Hui, X.; Zheng, S.; Hu, Y.; Xu, C.; Jin, X.; Chi, H.; Zhang, X. Ultralow Reflectivity Spiral Phase Plate for Generation of Millimeter-wave OAM Beam. *IEEE Antennas Wirel. Propag. Lett.* **2015**, *14*, 4876. [CrossRef]
9. Bai, Q.; Tennant, A.; Allen, B. Experimental Circular Phased Array for Generating OAM Radio Beams. *Electron. Lett.* **2014**, *50*, 1414–1415. [CrossRef]
10. Mohammadi, S.M.; Daldorff, L.K.S.; Bergman, J.E.S.; Karlsson, R.L.; Thide, B.; Forozesh, K.; Carozzi, T.D.; Isham, B. Orbital Angular Momentum in Radio-A System Study. *IEEE Trans. Antennas Propag.* **2010**, *58*, 565–572. [CrossRef]
11. Barbuto, M.; Trotta, F.; Bilotti, F.; Toscano, A. Circular Polarized Patch Antenna Generating Orbital Angular Momentum. *Prog. Electromagn. Res.* **2014**, *148*, 23–30. [CrossRef]
12. Ren, J.; Leung, K.W. Generation of microwave orbital angular momentum states using hemispherical dielectric resonator antenna. *Appl. Phys. Lett.* **2018**, *112*, 131103. [CrossRef]
13. Chavez-Cerda, S.; Padgett, M.J.; Allison, I.; New, G.H.C.; Gutierrez-Vega, J.C.; O'Neil, A.T.; MacVicar, I.; Courtial, J. Holographic generation and orbital angular momentum of high-order Mathieu beams. *J. Opt. B-Quantum Semicl. Opt.* **2002**, *4*, S52–S57. [CrossRef]
14. Feng, R.; Yi, J.J.; Burokur, S.N.; Kang, L.; Zhang, H.L.; Werner, D.H. Orbital angular momentum generation method based on transformation electromagnetics. *Opt. Express* **2018**, *26*, 11708–11717. [CrossRef] [PubMed]
15. Chen, M.L.N.; Jiang, L.J.; Sha, W.E.I. Artificial perfect electric conductor-perfect magnetic conductor anisotropic metasurface for generating orbital angular momentum of microwave with nearly perfect conversion efficiency. *J. Appl. Phys.* **2016**, *119*, 064506. [CrossRef]
16. Chen, M.; Jiang, L.J.; Sha, W.E.I. Ultrathin Complementary Metasurface for Orbital Angular Momentum Generation at Microwave Frequencies. *IEEE Trans. Antennas Propag.* **2017**, *65*, 396–400. [CrossRef]
17. Zhang, L.; Mei, S.T.; Huang, K.; Qiu, C.W. Advances in Full Control of Electromagnetic Waves with Metasurfaces. *Adv. Opt. Mater.* **2016**, *4*, 818–833. [CrossRef]
18. Yu, S.X.; Li, L.; Shi, G.M.; Zhu, C.; Zhou, X.X.; Shi, Y. Design, fabrication, and measurement of reflective metasurface for orbital angular momentum vortex wave in radio frequency domain. *Appl. Phys. Lett.* **2016**, *108*, 121903. [CrossRef]
19. Yu, S.X.; Li, L.; Shi, G.M. Dual-polarization and dual-mode orbital angular momentum radio vortex beam generated by using reflective metasurface. *Appl. Phys. Express* **2016**, *9*, 082202. [CrossRef]
20. Yang, J.; Zhang, C.; Ma, H.F.; Zhao, J.; Dai, J.Y.; Yuan, W.; Yang, L.X.; Cheng, Q.; Cui, T.J. Generation of radio vortex beams with designable polarization using anisotropic frequency selective surface. *Appl. Phys. Lett.* **2018**, *112*, 203501. [CrossRef]
21. Yu, S.X.; Li, L.; Shi, G.M.; Zhu, C.; Shi, Y. Generating multiple orbital angular momentum vortex beams using a metasurface in radio frequency domain. *Appl. Phys. Lett.* **2016**, *108*, 241901. [CrossRef]

22. Zhang, C.; Deng, L.; Zhu, J.; Hong, W.; Wang, L.; Yang, W.; Li, S. Control of the Spin Angular Momentum and Orbital Angular Momentum of a Reflected Wave by Multifunctional Graphene Metasurfaces. *Materials* **2018**, *11*, 1054. [CrossRef] [PubMed]
23. Wang, L.; Shi, H.; Zhu, S.; Li, J.; Zhang, A.; Li, L. Generation of multiple modes microwave vortex beams using tunable metasurface. In Proceedings of the 2017 7th IEEE International Symposium on Microwave, Antenna, Propagation, and EMC Technologies (MAPE), Xi'an, China, 24–27 October 2017; pp. 379–381.
24. Jiang, Z.H.; Kang, L.; Hong, W.; Werner, D.H. Highly Efficient Broadband Multiplexed Millimeter-Wave Vortices from Metasurface-Enabled Transmit-Arrays of Subwavelength Thickness. *Phys. Rev. Appl.* **2018**, *9*, 064009. [CrossRef]
25. Chen, M.L.N.; Jiang, L.J.; Sha, W.E.I. Orbital Angular Momentum Generation and Detection by Geometric-Phase Based Metasurfaces. *Appl. Sci.* **2018**, *8*, 362. [CrossRef]
26. Bliokh, K.Y.; Bekshaev, A.Y.; Nori, F. Dual electromagnetism: Helicity, spin, momentum and angular momentum. *New J. Phys.* **2013**, *15*, 033026. [CrossRef]
27. Liu, Y.J.; Xia, S.; Shi, H.Y.; Zhang, A.X.; Xu, Z. Dual-band and high-efficiency polarization converter based on metasurfaces at microwave frequencies. *Appl. Phys. B-Lasers Opt.* **2016**, *122*, 178. [CrossRef]
28. Dong, G.X.; Shi, H.Y.; Xia, S.; Li, W.; Zhang, A.X.; Xu, Z.; Wei, X.Y. Ultra-broadband and high-efficiency polarization conversion metasurface with multiple plasmon resonance modes. *Chin. Phys. B* **2016**, *25*, 084202. [CrossRef]
29. Ran, Y.Z.; Liang, J.G.; Cai, T.; Li, H.P. High-performance broadband vortex beam generator using reflective Pancharatnam-Berry metasurface. *Opt. Commun.* **2018**, *427*, 101–106. [CrossRef]

materials

MDPI

Article

High-Efficiency and Wide-Angle Versatile Polarization Controller Based on Metagratings

Kun Song [1,*,†], Ruonan Ji [1,†], Duman Shrestha [2], Changlin Ding [1], Yahong Liu [1], Weiren Zhu [3], Wentao He [1], Huidong Liu [1], Yuhua Guo [1], Yongkang Tang [1], Xiaopeng Zhao [1,*] and Jiangfeng Zhou [2,*]

1 Department of Applied Physics, Northwestern Polytechnical University, Xi'an 710129, China; jiruonan@nwpu.edu.cn (R.J.); dingchanglin@nwpu.edu.cn (C.D.); yhliu@nwpu.edu.cn (Y.L.); leimu666233@163.com (W.H.); a1261843107@126.com (H.L.); 15353506120@163.com (Y.G.); guada520@mail.nwpu.edu.cn (Y.T.)

2 Department of Physics, University of South Florida, 4202 East Fowler Ave, Tampa, FL 33620, USA; duman@mail.usf.edu

3 Department of Electronic Engineering, Shanghai Jiao Tong University, Shanghai 200240, China; weiren.zhu@sjtu.edu.cn

* Correspondence: songkun@nwpu.edu.cn (K.S.); xpzhao@nwpu.edu.cn (X.Z.); jiangfengz@usf.edu (J.Z.); Tel.: +86-029-8843-1662 (X.Z.)

† These authors contributed equally to this work.

Received: 28 January 2019; Accepted: 15 February 2019; Published: 19 February 2019

Abstract: Metamaterials with their customized properties enable us to efficiently manipulate the polarization states of electromagnetic waves with flexible approaches, which is of great significance in various realms. However, most current metamaterial-based polarization controllers can only realize single function, which has extremely hindered the expansion of their applications. Here, we experimentally demonstrate highly efficient and multifunctional polarization conversion effects using metagrating by integrating single-structure metallic meta-atoms into the dielectric gratings. Benefiting from the combined advantages of the gratings and the metamaterials, the considered metagrating can operate in transmission and reflection modes simultaneously, acting as a high-performance and wide-angle quarter-wave or half-wave plate with distinct functions in different frequency bands. This metagrating structure is scalable to other frequency ranges and may provide opportunities to design compact multifunctional optical polarization control devices.

Keywords: metagratings; polarization controller; multifunction; wide-angle; dual mode

1. Introduction

As one of the key properties of an electromagnetic wave, polarization has numerous fascinating applications in the scientific research area as well as our daily life. However, limited by the intrinsic electromagnetic properties of natural materials, the conventional polarization control devices often suffer from bulky configurations, low efficiencies, and narrow working bandwidths [1], which extremely hinders their potential applications. In addition, to comply with the trend of integration and miniaturization of the electromagnetic system, integration of multiple diversified functionalities into a single and compact device has attracted increasing attention.

In recent years, the development of metamaterials opened new opportunities to efficiently manipulate electromagnetic waves [2–6]. With well-designed meta-atoms, metamaterials exhibit an incomparable superiority in achieving giant anisotropy or chirality which is much larger than that of natural materials [7–11], making it possible to control electromagnetic wave polarizations with sub-wavelength profiles [7,10,12–14]. On this basis, various promising ultra-compact polarization controllers based on metamaterials have been proposed, such as polarization rotators [15–18], quarter-wave plates [19–22], half-wave plates [23–25], asymmetric transmission devices [26–28], and chiral mirrors [29–32]. Despite the great progress, most of the aforementioned metamaterial-based polarization controllers possess single functionality only, which inevitably restricts the application flexibilities to some degree. Recently, considerable efforts have been devoted to integrating multiple functionalities into a single meta-device [31,33–36], and some multifunctional polarization controllers, such as multi-modal reflective metasurface polarization generator [37], metasurface with absorption and polarization conversion functions [38], metasurface with reconfigurable conversions of reflection, transmission, and polarization states [39], have been demonstrated. Additionally, some previous papers have concluded that grating structures can also realize multiple polarization conversions [40–42]. In spite of the described advantages, some of these polarization controllers need to integrate multifarious complex supercells to accomplish multi-functionalities [37], which are unfavorable to the structure design and fabrication process. Additionally, limited by the current design concept, these multifunctional polarization controllers work in single reflective or transmissive mode merely [37–42], and to date, a multifunctional polarization controller that can operate in dual modes, which may find important applications in multipath systems, has not yet been obtained. Thus, efficiently integrating multiple distinct polarization conversion functions and dual operating modes into a single polarization controller composed of simple geometric structures is meaningful and still a challenge.

More recently, metagratings with excellent performances have been proposed to efficiently manipulate electromagnetic waves. For instance, Khorasaninejad et al. demonstrated that a metagrating composed of dielectric ridge waveguides can realize broadband and efficient routing (splitting and bending) into a single diffraction order and additional polarization beam splitter capabilities [43]. In another study, Ra'di et al. proposed a metagrating consisting of periodic arrays of anisotropic inclusions to control the wave front with unitary efficiency [44]. However, manipulating the polarization state of the electromagnetic waves via metagratings has not yet been reported. Inspired by the concept of metagratings, in this paper we propose a new route to manipulate the polarization state and amplitude of electromagnetic waves with an anisotropic metagrating which is obtained through the combination of metallic metamaterial and dielectric grating. Furthermore, by delicately designing the anisotropic meta-atoms, the unique metagrating is capable of realizing polarization-sensitive phase responses and can efficiently work in the transmission and reflection modes simultaneously. Both of the simulation and experimental results showed that multiple high-efficiency diversified functionalities, including linear polarization rotator, circular polarizer, circular polarization converter, and chirality preserving mirror, are integrated into this single sub-wavelength meta-device. Moreover, the corresponding efficient polarization conversion behaviors in different frequency bands are all insensitive to the incident angle of electromagnetic waves. Such a multifunctional metagrating will provide greater flexibility in a variety of practical applications where polarization controllers are involved.

2. Theory and Structure Design

When an anisotropic structure with the principal axis along u- and v-axis directions, for instance, a typical metal grating shown in Figure 1a, is illuminated by a plane wave propagating along w-axis with the polarization orientation of 45° with respect to u-axis direction, the reflected and transmitted electric fields can be written as:

$$\begin{pmatrix} E_{ou} \\ E_{ov} \end{pmatrix} = \begin{pmatrix} a & 0 \\ 0 & b \end{pmatrix} \begin{pmatrix} E_{iu} \\ E_{iv} \end{pmatrix} = a \begin{pmatrix} E_{iu} \\ \alpha Exp(i\beta) E_{iv} \end{pmatrix}, \tag{1}$$

where a and b are the reflection and transmission coefficients of the anisotropic structure along u- and v-axis, respectively, and the ratio of $\alpha = |b|/|a|$ is defined as the amplitude ratio. β is the phase difference between the two principal axes. For a perfect polarization converter, the value of α should be equal to 1 and the value of β should be $\pm\pi/2$ for the quarter-wave plate and $\pm\pi$ for the half-wave plate.

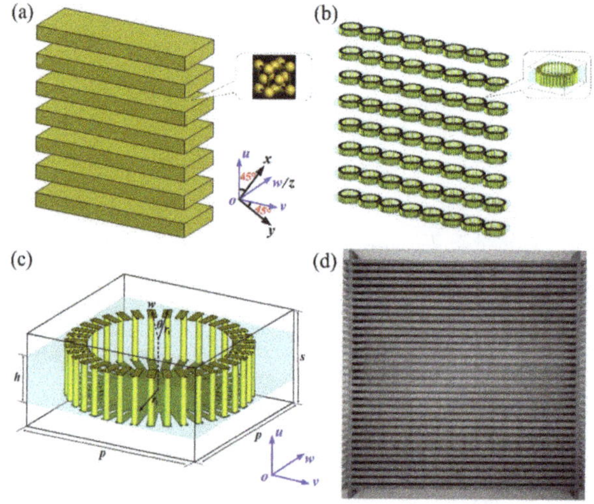

Figure 1. Schematic diagram of (**a**) the metallic subwavelength grating, (**b**) the proposed metagrating, and (**c**) the unit cell of metagrating. The u- and v-axis are the two principal axes of the metagrating, respectively. The direction of the $x(y)$-axis is rotated by 45° with respect to the $u(v)$-axis. (**d**) Photograph of the experimental sample.

Due to the dispersion, β may achieve $\pm\pi/2$ and $\pm\pi$ at different wavelengths, thereby a single device can exhibit multifunctional polarization conversion as both quarter-wave and half-wave plates [40,41]. However, such multifunctional wave plates are only able to operate within a narrow band due to the dispersion nature.

In this work, we choose the grating structure to achieve multiple polarization conversion because of its simultaneous control of phases and amplitudes capabilities [40–42,45]. Benefiting from the concept of metamaterials, we are able to control α, β, and even operating modes through a more flexible approach by integrating artificial meta-atoms into conventional gratings. Here, as shown in Figure 1b, we propose a sub-wavelength metagrating composed of single wheel-like unit cells as an example. Figure 1c exhibits the detailed geometrical structure of the unit cell. The bilayer metallic patterns are connected by a metallization hole with the diameter of 0.6 mm. The structural parameters of the unit cell are as follows: p = 12 mm, s = 6 mm, r_1 = 5.9 mm, r_2 = 3.5mm, w = 0.5 mm, h = 3 mm, and θ = 10°. The metal cladding is copper with the thickness of 0.035 mm and conductivity of $\sigma = 5.8 \times 10^7$ S/m. The dielectric substrate is Teflon with a relative permittivity of $2.2 + 0.001 \times i$. Figure 1d shows the photograph of the fabricated metagrating structure, which is composed of 33 layers of slats each consisting of 16 wheel-like unit cells.

3. Results and Discussion

In order to get the electromagnetic properties of the proposed metagrating, both of the simulations and experiments were carried out. The simulations were achieved with the commercial finite element software CST Microwave Studio (CST2018, Computer Simulation Technology GmbH). In the simulations, unit cell boundary conditions were used in the u- and v-axis directions, while open boundary conditions were employed in the w-axis direction. The experimental data was obtained using a network analyzer (AV3629, 41st institute of CETC, Qingdao, China) with two broadband linearly polarized horn antennas in an anechoic chamber.

Since the considered metagrating can work in the transmission and reflection modes at the same time, we first studied the transmission mode of the metagrating in the frequency range of 5.5 GHz to 8.0 GHz. Figure 2 shows the transmittance, amplitude ratio, and phase difference of the proposed metagrating as the linearly polarized incident waves are u- and v-polarizations. In this paper, the first subscript m in T_{mn}/R_{mn} represents the polarization state of the transmitted and reflected wave, while the second one n represents the incident wave. As shown in Figure 2a, the co-polarization spectrum (T_{vv}) of v-polarization incident waves shows there are three peaks with high transmittance occurring on the co-polarization spectrum (T_{vv}) of v-polarization incident waves. For the u-polarization incident waves, the metagrating will operate similarly as a regular wire-grid metallic grating, which enables a high transmittance for the linearly incident waves with the polarization plane perpendicular to its wire line direction [18,46], leading to a high co-polarization transmittance (T_{uu}) over 0.95 in the whole frequency range. Since the transmittance of T_{uu} and T_{vv}, are close to each other around the frequencies of 6.1 GHz, 7.1 GHz, and 7.9 GHz, the amplitude ratios $\alpha_T = T_{vv}/T_{uu}$ in Figure 2c are close to 1 around the corresponding frequencies, which is important to ensure a perfect quarter-wave or half-wave plate performance. Figure 2e portrays the transmission phase difference β_T between u- and v-axis ($\beta_T = \varphi_{vv} - \varphi_{uu}$). It is obvious that the phase difference β_T can be varied in a large dynamic range covering from 0 to 2π. And it can also be found that around the frequencies of 5.8 GHz, 6.4 GHz, 7.1 GHz, and 7.9 GHz, the phase differences are approximately about $\pi/2$, π, $3\pi/2$, and $\pi/2$, respectively, which implies that the metagrating is able to function as a transmission-type half-wave or quarter-wave plate at the selected frequencies. The experimental results of the proposed metagrating are shown in Figure 2b,d,f, respectively, which show good agreement with simulations.

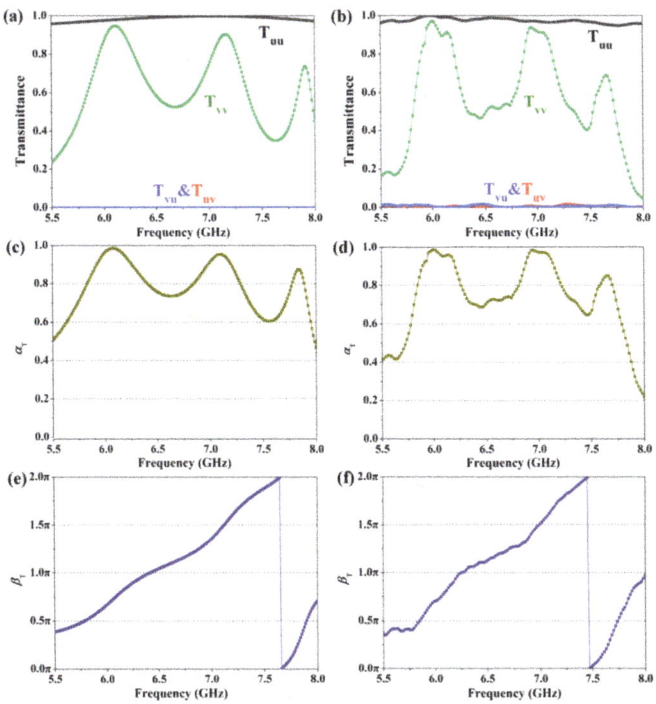

Figure 2. Simulation (left column) and experimental (right column) results of the proposed metagrating in the case of u- and v-polarization incidence. In this frequency region, the metamaterial operates in transmission mode. (**a**,**b**) Transmittance, (**c**,**d**) calculated amplitude ratio α_T, (**e**,**f**) phase difference β_T between u- and v-axis.

In Figure 3, we illustrate the results of the considered metagrating in the case of x- and y-polarized wave incidence. It is significant that in Figure 3a the co-polarization and cross-polarization transmission spectra for the x- and y-polarized incident waves are completely coincident, i.e., $T_{xx} = T_{yy}$ and $T_{yx} = T_{xy}$. This phenomenon can be attributed to the unit cell that is mirror symmetry with respect to the plane $x + y = 0$ (uow plane) [23]. In the frequency range from 5.8 GHz to 7.1 GHz, the cross-polarized waves account for the majority of transmitted waves, and the highest transmittance of the cross-polarization rise up to 0.87 at 6.2 GHz, which means that as an $x(y)$-polarized incident wave passes through the proposed metagrating, the transmitted wave is mainly transformed into $y(x)$-polarized wave around this frequency. While in the frequency range of 7.1 GHz to 7.9 GHz, the co-polarized waves dominate the transmitted waves, and the highest transmittance is 0.75 at 7.8 GHz. As for the intersections of co-polarization and cross-polarization transmission curves, the metagrating will work as a quarter-wave plate. According to the phase differences β_T shown in Figure 2e,f, in the frequency vicinity of 5.8 GHz, 7.1 GHz, and 7.9 GHz, the transmitted waves are left-handed (right-handed), right-handed (left-handed), and left-handed (right-handed) circular polarizations with the transmittance over 0.78 for an $x(y)$-polarized incident wave, respectively. Particularly, the transmittance of the right-handed circularly polarization wave is nearly close to 1 at 7.1 GHz, indicating a perfect linear-to-circular polarization conversion effect. In Figure 3c, the polarization conversion ratio (PCR), which is defined as PCR = $T_{cross}/(T_{cross} + T_{co})$, is used to characterize the polarization conversion efficiency of the metagrating. It shows that the PCR is larger than 0.9 between 6.1 GHz and 6.8 GHz and reaches its maximum value of 0.99 at 6.4 GHz, where the metagrating will operate as a nearly perfect half-wave plate. The experimentally measured results are respectively plotted in Figure 3b,d, which exhibit similar phenomena as the simulations in spite of the slight deviations. According to the aforementioned results, the proposed metagrating is capable of acting as a high-performance multiband transmission-type wave plate.

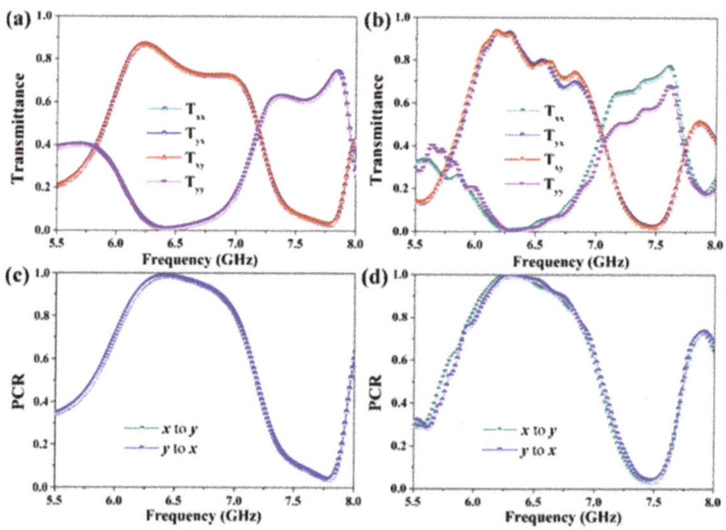

Figure 3. Simulated (left column) and measured (right column) results of the designed metagrating for the x- and y-polarized incident waves in transmission mode. (**a**,**b**) Transmittance, (**c**,**d**) PCR.

Between 10.0 GHz and 15.0 GHz, the intriguing metagrating will operate in reflective mode. Figure 4 plots the results of the proposed metagrating for u- and v-polarized wave incidence. In Figure 4a,b, it is seen that the metagrating can realize high reflection for both of u- and v-polarized incident waves, with the simulated (experimental) reflectance of R_{uu} and R_{vv} 0.85 (0.83) from 10.0 GHz to 14.5 GHz. The corresponding results of the amplitude ratios α_R are shown in Figure 4c,d, separately.

Obviously, the values of α_R are approximately 1 in this frequency range. Additionally, it can be found from Figure 4e,f, that the phase differences of reflection waves along the two principal axes are roughly maintained π in the frequency range of 10.0 GHz to 14.5 GHz. And the features mentioned above enable the metagrating to act as a broadband reflection-type half-wave plate with high efficiency.

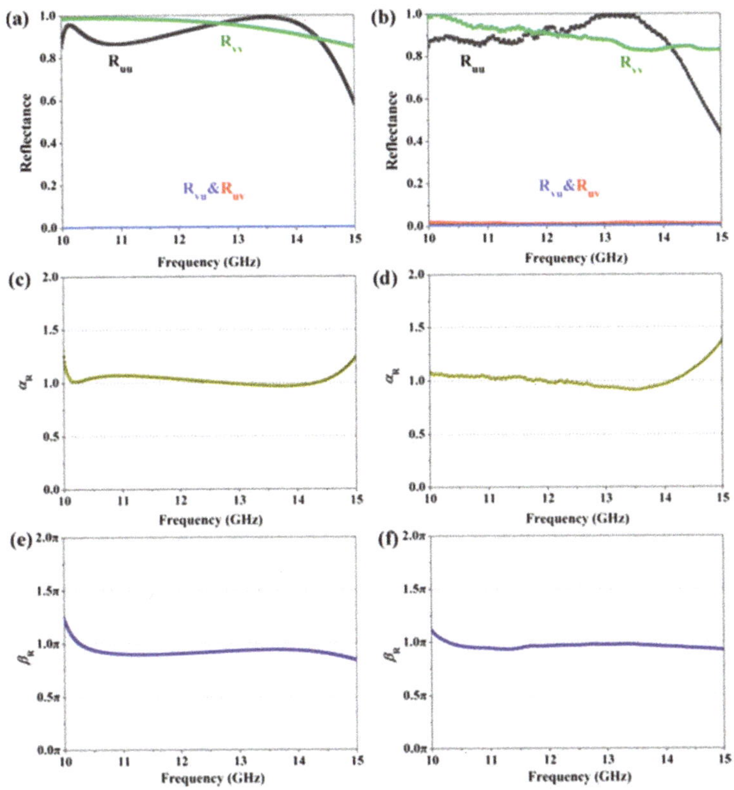

Figure 4. Numerical (left column) and experimental (right column) results of the proposed metamaterial under *u*- and *v*-polarization incidence in the reflective frequency range. (**a**,**b**) Reflectance, (**c**,**d**) amplitude ratio α_R, (**e**,**f**) phase difference β_R between *u*- and *v*-axis.

Figure 5 shows the reflective spectra and the corresponding PCR of the designed metagrating for the *x*- and *y*-polarized incident waves. As shown in Figure 5a,b, the co-polarization and cross-polarization reflection spectra for the *x*-polarized incident wave are as same as those of *y*-polarized incident wave, respectively. In the frequency range of 10.0 GHz to 14.3 GHz (the relative bandwidth is about 35.4%), the reflectance of cross-polarization reflection spectra, R_{yx} and R_{xy}, are larger than 0.9, while those of the co-polarization reflection spectra, R_{xx} and R_{yy}, are less than 0.03. This fact reveals that, for an *x*(*y*)-polarized incident wave, the reflected wave will be efficiently converted to be *y*(*x*)-polarized wave. In Figure 5c,d, we further calculate the PCR of the reflection waves of the metagrating. It is significant that both of the simulated and measured PCR are over 0.95 from 10.0 GHz to 14.3 GHz, exhibiting a nearly perfect reflective cross-polarization conversion effect. Therefore, the considered metagrating is able to work as a high-performance reflection-type half-wave plate in a broad bandwidth.

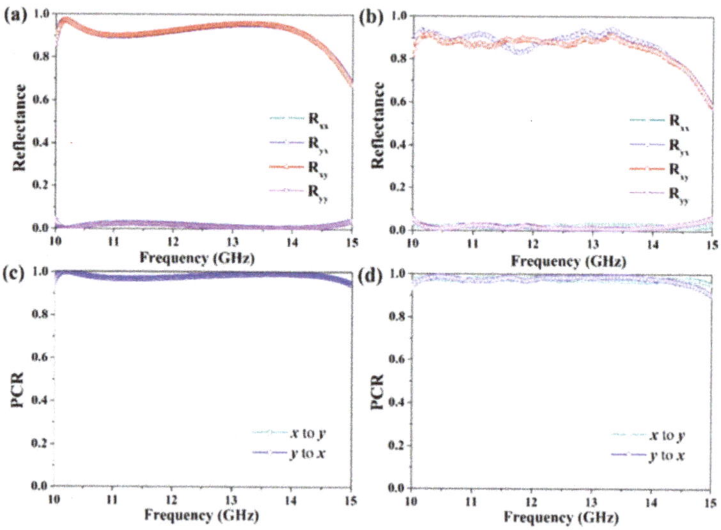

Figure 5. Simulation (left column) and experimental (right column) results of the proposed metagrating for the x- and y-polarized incident waves in the reflective mode. (**a**,**b**) Reflectance, (**c**,**d**) PCR.

The electromagnetic properties of the metagrating for the linearly polarized waves at oblique incidence were also studied, as shown in Figure 6. Figure 6a shows the influences of incident angle on the cross-polarization spectra, co-polarization spectra, and PCR of the metagrating in transmission mode for the x-polarized incident wave. Owing to the isotropic design of the unit cell, our metagrating shows consistent performance as the incident angle varies within a large range. As the incident angle increases, the transmittance of co-polarization and cross-polarization transmission at 5.8 GHz and 7.9 GHz both slightly increase, while a slight reduction phenomenon occurs at 7.1 GHz. Despite all this, the considered metagrating can still work well as an excellent quarter-wave plate with high transmittance over 0.8 at these frequencies even if the incident angle rises up to 40°. Additionally, it is worth noting that the variation of incident angle has almost no effect on the maximum transmittance at 6.2 GHz, and the corresponding PCR remain as high as 0.95 when the incident angle is 60°, implying a highly efficient and wide-angle half-wave plate function. In Figure 6b, it is seen that the co-polarized reflectance increases slightly as the incident angle increases, while the cross-polarized reflectance gradually decreases. In spite of the decrement, the cross-polarized reflectance is still larger than 0.8 from 10 GHz to 13.8 GHz as the incident angle is 40°, simultaneously accompanied by a PCR over 0.9. Hence, the broadband and efficient reflective cross-polarization conversion effect of the metagrating is insensitive to the incident angle of the x-polarized wave. In the case of y-polarization incidence, a similar polarization conversion of the transmission can also be observed, although the transmittance of the right-handed circularly polarized wave was decreased at around 5.8 GHz. More interestingly, the reflection mode of the metagrating exhibits a much stronger cross-polarization conversion with a wider bandwidth and a lager range of angle invariance in comparison to the case of x-polarization incidence, as shown in Figure 6c,d. Thus, the intriguing metagrating can function as a dual operating mode, high-performance, and multiple functional wave plate regardless of the incident angle, which exhibits more flexibility than the previous metamaterial-based wave plates [15–17,23,25].

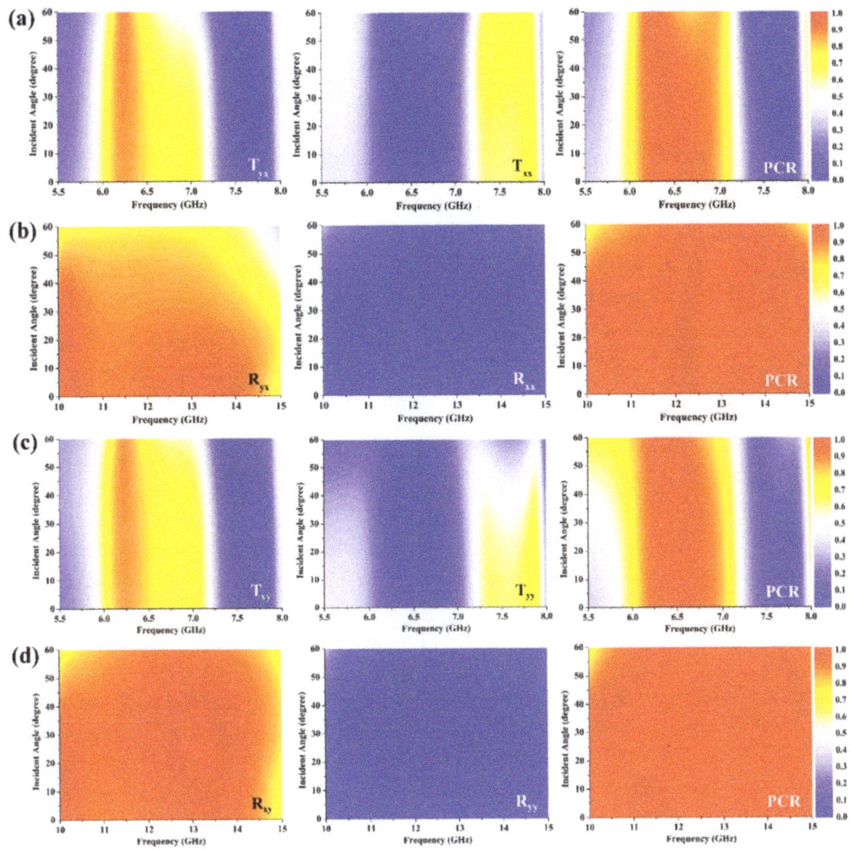

Figure 6. The effects of incident angle on the electromagnetic properties of the metagrating for the linearly polarized incident waves. The results of the metagrating in (**a**) transmission and (**b**) reflection modes under x-polarization incidence. The results of the metagrating in (**c**) transmission and (**d**) reflection modes under y-polarization incidence. At oblique incidence, the x- and y-polarization waves actually represent the TM and TE waves, respectively. In the simulations, the incident angle is increasingly tuned by a step of 5°.

As the designed metagrating is anisotropic, it can also operate for the circularly polarized wave incidence indeed. For the circular polarization basis, the circularly polarized transmission and reflection coefficients can be respectively obtained via the linear ones by the following equations [2,30]:

$$\begin{pmatrix} t_{++} & t_{+-} \\ t_{-+} & t_{--} \end{pmatrix} = \frac{1}{2}\begin{pmatrix} t_{xx}+t_{yy}+i(t_{xy}-t_{yx}) & t_{xx}-t_{yy}-i(t_{xy}+t_{yx}) \\ t_{xx}-t_{yy}+i(t_{xy}+t_{yx}) & t_{xx}+t_{yy}-i(t_{xy}-t_{yx}) \end{pmatrix} \tag{2}$$

$$\begin{pmatrix} r_{++} & r_{+-} \\ r_{-+} & r_{--} \end{pmatrix} = \frac{1}{2}\begin{pmatrix} r_{xx}+r_{yy}+i(r_{xy}-r_{yx}) & r_{xx}-r_{yy}-i(r_{xy}+r_{yx}) \\ r_{xx}-r_{yy}+i(r_{xy}+r_{yx}) & r_{xx}+r_{yy}-i(r_{xy}-r_{yx}) \end{pmatrix} \tag{3}$$

Here, the subscripts '+' and '–' represent the clockwise and counterclockwise circularly polarized waves when separately observed along +z direction. As the wave vectors are in opposite directions for transmitted and reflected circularly polarized waves, one should notice that each reflection component of the circularly polarized waves has different physical meaning compared with the transmission case. Thus, the transmission coefficients are defined as $t_{RR} = t_{++}$, $t_{LL} = t_{--}$, $t_{LR} = t_{-+}$, $t_{RL} = t_{+-}$, while the reflection coefficients are defined as $r_{LR} = r_{++}, r_{RR} = r_{-+}, r_{LL} = r_{+-}, r_{RL} = r_{--}$. Additionally, t_{RR}, t_{RL}, r_{RR}, and r_{RL} represent the right-handed circularly polarized (RCP) waves, while t_{LL}, t_{LR}, r_{LL}, and r_{LR} indicate the left-handed circularly polarized (LCP) waves.

According to the transmittance in Figure 3a and the reflectance in Figure 5a, we can easily deduce that the transmission and reflection coefficients satisfy the relationships of $t_{xx} = t_{yy}, t_{yx} = t_{xy}$, $r_{xx} = r_{yy}$, and $r_{yx} = r_{xy}$ at normal incidence, thus the Equations (2) and (3) can be further simplified as:

$$\begin{pmatrix} t_{++} & t_{+-} \\ t_{-+} & t_{--} \end{pmatrix} = \begin{pmatrix} t_{xx} & -it_{xy} \\ it_{xy} & t_{xx} \end{pmatrix}, \tag{4}$$

$$\begin{pmatrix} r_{++} & r_{+-} \\ r_{-+} & r_{--} \end{pmatrix} = \begin{pmatrix} r_{xx} & -ir_{xy} \\ ir_{xy} & r_{xx} \end{pmatrix}. \tag{5}$$

Namely, the transmittance and reflectance of the circularly polarized wave satisfy $T_{LL} = T_{RR} = T_{xx},\ T_{RL} = T_{LR} = T_{yx},\ R_{LL} = R_{RR} = R_{xy},\ R_{RL} = R_{LR} = R_{xx}$.

In Figure 7, we illustrate the simulated results of the metagrating for the circularly polarized waves at normal incidence. It is obvious that the transmittance and PCR spectra for the circularly polarized incident waves shown in Figure 7a,b, respectively, are completely the same as the ones in Figure 3. These phenomena can be well explained by Equations (2) and (4). It should be noted that, for the LCP (RCP) incident waves, the transmitted waves are $x(y)$-polarized, RCP (LCP), $y(x)$-polarized, and $x(y)$-polarized at 5.8 GHz, 6.4 GHz, 7.1 GHz, and 7.9 GHz, respectively. Figure 7c plots the reflectance spectra of the proposed metagrating. It can be found that as a circularly polarized wave is incident on the metagrating, the co-polarization reflectance (R_{LL} or R_{RR}) is larger than 0.9 from 10.1 GHz to 14.3 GHz, while the cross-polarization reflectance (R_{RL} or R_{LR}) is less than 0.03. That is to say, our design can function as a nearly perfect chirality preserving mirror in a broad bandwidth, consequently leading to a very low PCR of less than 0.03, as shown in Figure 7d.

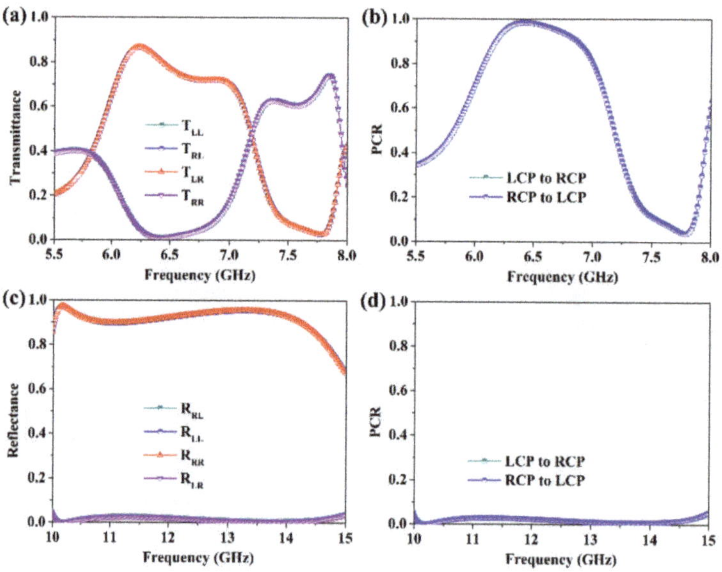

Figure 7. Simulation results of the designed metagrating for the circularly polarized waves at normal incidence. (**a**) Transmittance and (**b**) PCR spectra in transmission mode, (**c**) reflectance and (**d**) PCR spectra in reflection mode.

Figure 8 shows the influences of incident angle on the electromagnetic properties of the proposed metagrating in the case of circular polarization incidence. It can be seen in Figure 8a–c, that in transmission mode the operating frequencies generate slight blue shifts as the incident angle increases. Although the values of the cross-polarization transmittance gradually reduce as the incident angle increases, the highest transmittance of the cross-polarization can still be maintained over about 0.8 when the incident angle increases to 40°, simultaneously accompanied by a high PCR larger than 0.9. In Figure 8d–f, it is shown that the effects of incident angle on the co-polarization and cross-polarization reflectance spectra are inappreciable. And the lowest value of co-polarization reflectance is still over about 0.9 in the frequency range of 10.0 GHz to 14.3 GHz even when the incident angle is 40°, while the cross-polarization reflectance is no more than 0.05 with the corresponding PCR being less than 0.05. The aforementioned results further confirm that the proposed metagrating can also act as a dual-mode, wide-angle, and multifunctional wave plate with high-efficiency for the circularly polarized incident waves.

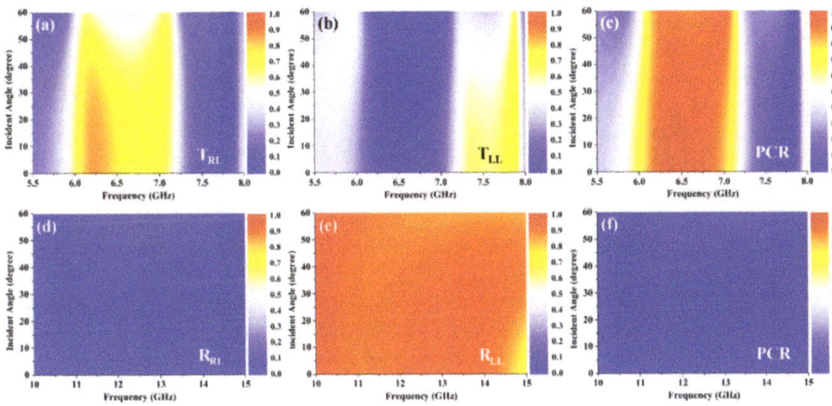

Figure 8. Influences of incident angle on the electromagnetic properties of the metagrating in the case of circular polarization incidence. (**a**) Cross-polarization transmittance, (**b**) co-polarization transmittance, and (**c**) PCR in transmission mode; (**d**) cross-polarization reflectance, (**e**) co-polarization reflectance, and (**f**) PCR in reflection mode.

4. Conclusions

In summary, we have proposed an effective way to achieve multifunctional polarization manipulation by a metagrating consisting of periodically arranged meta-atoms inside dielectric gratings. As a proof-of-concept experiment, we have designed and fabricated a sub-wavelength metagrating that consists of wheel-like meta-atoms. The metagrating works in both transmission and reflection modes efficiently. The phase differences between the two principal axes are monotonically altering with the frequencies in transmission mode covering a large dynamic range from 0 to 2π, while they are approximately kept at about π in a broad bandwidth region in reflective mode. Our measurements show that the metagrating can work as a dual-mode, high-efficiency, and multifunctional wave plate to realize various functions including linear-to-circular polarization conversion, linear or circular cross-polarization conversion, and chirality preserving mirror in different frequency bands simultaneously accompanied by a large angular invariance. Such versatile functionalities provide great flexibilities for optical polarization control devices. The design of metagrating can be extended to other frequency ranges as a compact optical polarization controller in the applications of telecommunications, radar detections, and optical devices.

Author Contributions: Conceptualization and methodology, K.S., R.J., J.Z. and X.Z.; data analysis, K.S., R.J., W.Z. and J.Z.; investigation, C.D., W.H., H.L., Y.G., Y.T. and Y.L.; software, K.S., R.J., D.S., C.D. and W.Z.; writing—original draft preparation, K.S., R.J. and J.Z.; writing—review and editing, K.S., R.J., J.Z. and X.Z.

Funding: This research was funded by the National Natural Science Foundation of China (Grant Nos. 61601375, 61805204, 11874301, 11674267 and 61701303); Natural Science Basic Research Plan in Shaanxi Province of China (Grant Nos. 2018JQ1036 and 2017JM1009); Fundamental Research Funds for the Central Universities (Grant Nos. 3102016ZY029, 3102017OQD076, 3102017jghk02004 and 3102017zy015); KRISS grant GP2018-0023; AOARD grant FA2386-18-1-4104 funded by the U.S. government (AFOSR/AOARD).

Conflicts of Interest: The authors declare no conflict of interest.

References

1. Born, M.; Wolf, E. *Principles of Optics*, 7th ed.; Cambridge University Press: Cambridge, UK, 1999; pp. 109–132.
2. Zhou, J.; Dong, J.; Wang, B.; Koschny, T.; Kafesaki, M.; Soukoulis, C.M. Negative refractive index due to chirality. *Phys. Rev. B* **2009**, *79*, 121104. [CrossRef]
3. Zhang, S.; Zhou, J.; Park, Y.S.; Rho, J.; Singh, R.; Nam, S.; Azad, A.K.; Chen, H.T.; Yin, X.; Taylor, A.J.; et al. Photoinduced handedness switching in terahertz chiral metamolecules. *Nat. Commun.* **2012**, *3*, 942. [CrossRef] [PubMed]
4. Shi, H.; Wang, L.; Zhao, M.; Chen, J.; Zhang, A.; Xu, Z. Transparent metasurface for generating microwave vortex beams with cross-polarization conversion. *Materials* **2018**, *11*, 2448. [CrossRef] [PubMed]
5. Qin, F.; Ding, L.; Zhang, L.; Monticone, F.; Chum, C.C.; Deng, J.; Mei, S.; Li, Y.; Teng, J.; Hong, M.; et al. Hybrid bilayer plasmonic metasurface efficiently manipulates visible light. *Sci. Adv.* **2016**, *2*, e1501168. [CrossRef] [PubMed]
6. Song, K.; Ding, C.; Su, Z.; Liu, Y.; Luo, C.; Zhao, X.; Bhattarai, K.; Zhou, J. Planar composite chiral metamaterial with broadband dispersionless polarization rotation and high transmission. *J. Appl. Phys.* **2016**, *120*, 245102. [CrossRef]
7. Hao, J.; Yuan, Y.; Ran, L.; Jiang, T.; Kong, J.; Chan, C.; Zhou, L. Manipulating electromagnetic wave polarizations by anisotropic metamaterials. *Phys. Rev. Lett.* **2007**, *99*, 063908. [CrossRef] [PubMed]
8. Zhou, J.; Chowdhury, D.R.; Zhao, R.; Azad, A.K.; Chen, H.; Soukoulis, C.M.; Taylor, A.J.; O'Hara, J.F. Terahertz chiral metamaterials with giant and dynamically tunable optical activity. *Phys. Rev. B* **2012**, *86*, 035448. [CrossRef]
9. Yang, J.; Cheng, Y.; Ge, C.; Gong, R. Broadband polarization conversion metasurface based on metal cut-wire structure for radar cross section reduction. *Materials* **2018**, *11*, 626. [CrossRef]
10. Baimuratov, A.S.; Pereziabova, T.P.; Zhu, W.; Leonov, M.Y.; Baranov, A.V.; Fedorov, A.V.; Rukhlenko, I.D. Optical anisotropy of topologically distorted semiconductor nanocrystals. *Nano Lett.* **2017**, *17*, 5514–5520. [CrossRef]
11. Song, K.; Liu, Y.; Fu, Q.; Zhao, X.; Luo, C.; Zhu, W. 90° polarization rotator with rotation angle independent of substrate permittivity and incident angles using a composite chiral metamaterial. *Opt. Express* **2013**, *21*, 7439–7446. [CrossRef]
12. Zhao, R.; Zhang, L.; Zhou, J.; Koschny, T.; Soukoulis, C.M. Conjugated gammadion chiral metamaterial with uniaxial optical activity and negative refractive index. *Phys. Rev. B* **2011**, *83*, 035105. [CrossRef]
13. Deng, L.; Zhang, Y.; Zhu, J.; Zhang, C. Wide-band circularly polarized reflectarray using graphene-based Pancharatnam-Berry phase unit-cells for terahertz communication. *Materials* **2018**, *11*, 956. [CrossRef] [PubMed]
14. Song, K.; Su, Z.; Wang, M.; Silva, S.; Bhattarai, K.; Ding, C.; Liu, Y.; Luo, C.; Zhao, X.; Zhou, J. Broadband angle- and permittivity-insensitive nondispersive optical activity based on planar chiral metamaterials. *Sci. Rep.* **2017**, *7*, 10730. [CrossRef] [PubMed]
15. Ye, Y.; He, S. 90° polarization rotator using a bilayered chiral metamaterial with giant optical activity. *Appl. Phys. Lett.* **2010**, *96*, 203501. [CrossRef]
16. Mutlu, M.; Ozbay, E. A transparent 90° polarization rotator by combining chirality and electromagnetic wave tunneling. *Appl. Phys. Lett.* **2012**, *100*, 051909. [CrossRef]
17. Song, K.; Zhao, X.; Liu, Y.; Fu, Q.; Luo, C. A frequency-tunable 90°-polarization rotation device using composite chiral metamaterials. *Appl. Phys. Lett.* **2013**, *103*, 101908. [CrossRef]

18. Fan, R.; Zhou, Y.; Ren, X.; Peng, R.; Jiang, S.; Xu, D.; Xiong, X.; Huang, X.; Wang, M. Freely tunable broadband polarization rotator for terahertz waves. *Adv. Mater.* **2015**, *27*, 1201–1206. [CrossRef] [PubMed]
19. Zhao, Y.; Belkin, M.A.; Alù, A. Twisted optical metamaterials for planarized ultrathin broadband circular polarizers. *Nat. Commun.* **2012**, *3*, 870. [CrossRef] [PubMed]
20. Yu, N.; Aieta, F.; Genevet, P.; Kats, M.A.; Gaburro, Z.; Capasso, F. A broadband, background-free quarterwave plate based on plasmonic metasurfaces. *Nano Lett.* **2012**, *12*, 6328–6333. [CrossRef] [PubMed]
21. Jiang, S.; Xiong, X.; Hu, Y.; Hu, Y.; Ma, G.; Peng, R.; Sun, C.; Wang, M. Controlling the polarization state of light with a dispersion-free metastructure. *Phys. Rev. X* **2014**, *4*, 021026. [CrossRef]
22. Ji, R.; Wang, S.; Liu, X.; Guo, H.; Lu, W. Hybrid helix metamaterials for giant and ultrawide circular dichroism. *ACS Photonics* **2016**, *3*, 2368–2374. [CrossRef]
23. Wu, X.; Meng, Y.; Wang, L.; Tian, J.; Dai, S.; Wen, W. Anisotropic metasurface with near-unity circular polarization conversion. *Appl. Phys. Lett.* **2016**, *108*, 183502. [CrossRef]
24. Liu, Z.; Li, Z.; Liu, Z.; Cheng, H.; Liu, W.; Tang, C.; Gu, C.; Li, J.; Chen, H.; Chen, S.; et al. Single-layer plasmonic metasurface half-wave plates with wavelength-independent polarization conversion angle. *ACS Photonics* **2017**, *4*, 2061–2069. [CrossRef]
25. Song, K.; Su, Z.; Silva, S.; Fowler, C.; Ding, C.; Ji, R.; Liu, Y.; Zhao, X.; Zhou, J. Broadband and high-efficiency transmissive-type nondispersive polarization conversion meta-device. *Opt. Mater. Express* **2018**, *8*, 2430–2438. [CrossRef]
26. Rogacheva, A.V.; Fedotov, V.A.; Schwanecke, A.S.; Zheludev, N.I. Asymmetric propagation of electromagnetic waves through a planar chiral structure. *Phys. Rev. Lett.* **2006**, *97*, 177401. [CrossRef] [PubMed]
27. Mutlu, M.; Akosman, A.E.; Serebryannikov, A.E.; Ozbay, E. Asymmetric transmission of linearly polarized waves and polarization angle dependent wave rotation using a chiral metamaterial. *Opt. Express* **2011**, *19*, 14290–14299. [CrossRef] [PubMed]
28. Song, K.; Liu, Y.; Luo, C.; Zhao, X. High-efficiency broadband and multiband cross-polarization conversion using chiral metamaterial. *J. Phys. D Appl. Phys.* **2014**, *47*, 505104. [CrossRef]
29. Plum, E.; Zheludev, N.I. Chiral mirrors. *Appl. Phys. Lett.* **2015**, *106*, 221901. [CrossRef]
30. Wang, Z.; Jia, H.; Yao, K.; Cai, W.; Chen, H.; Liu, Y. Circular dichroism metamirrors with near-perfect extinction. *ACS Photonics* **2016**, *3*, 2096–2101. [CrossRef]
31. Jing, L.; Wang, Z.; Maturi, R.; Zheng, B.; Wang, H.; Yang, Y.; Shen, L.; Hao, R.; Yin, W.; Li, E.; et al. Gradient chiral metamirrors for spin-selective anomalous reflection. *Laser Photonics Rev.* **2017**, *11*, 1700115. [CrossRef]
32. Kang, L.; Rodrigues, S.P.; Taghinejad, M.; Lan, S.; Lee, K.T.; Liu, Y.; Werner, D.H.; Urbas, A.; Cai, W. Preserving spin states upon reflection: Linear and nonlinear responses of a chiral meta-mirror. *Nano Lett.* **2017**, *17*, 7102–7109. [CrossRef] [PubMed]
33. Cai, T.; Tang, S.; Wang, G.; Xu, H.; Sun, S.; He, Q.; Zhou, L. High-performance bifunctional metasurfaces in transmission and reflection geometries. *Adv. Opt. Mater.* **2017**, *5*, 1600506. [CrossRef]
34. Xie, X.; Li, X.; Pu, M.; Ma, X.; Liu, K.; Guo, Y.; Luo, X. Plasmonic metasurfaces for simultaneous thermal infrared invisibility and holographic illusion. *Adv. Funct. Mater.* **2018**, *28*, 1706673. [CrossRef]
35. Li, Y.; Cai, B.; Cheng, Q.; Cui, T. Isotropic holographic metasurfaces for dual-functional radiations without mutual interferences. *Adv. Funct. Mater.* **2016**, *26*, 29–35. [CrossRef]
36. Cong, L.; Pitchappa, P.; Wu, Y.; Ke, L.; Lee, C.; Singh, N.; Yang, H.; Singh, R. Active multifunctional microelectromechanical system metadevices: Applications in polarization control, wavefront deflection, and holograms. *Adv. Opt. Mater.* **2017**, *5*, 1600716. [CrossRef]
37. Wu, P.C.; Tsai, W.Y.; Chen, W.T.; Huang, Y.W.; Chen, T.Y.; Chen, J.W.; Liao, C.Y.; Chu, C.H.; Sun, G.; Tsai, D.P. Versatile polarization generation with an aluminum plasmonic metasurface. *Nano Lett.* **2016**, *17*, 445–452. [CrossRef] [PubMed]
38. Cheng, H.; Wei, X.; Yu, P.; Li, Z.; Liu, Z.; Li, J.; Chen, S.; Tian, J. Integrating polarization conversion and nearly perfect absorption with multifunctional metasurfaces. *Appl. Phys. Lett.* **2017**, *110*, 171903. [CrossRef]
39. Tao, Z.; Wan, X.; Pan, B.; Cui, T. Reconfigurable conversions of reflection, transmission, and polarization states using active metasurface. *Appl. Phys. Lett.* **2017**, *110*, 121901. [CrossRef]
40. Hu, J.; Lin, Y.; Zhao, X.; Zhu, A.; Xu, C.; Zhu, X.; Guo, P.; Cao, B.; Wang, C. All-metal flexible large-area multiband waveplate. *Opt. Express* **2017**, *25*, 8245–8254. [CrossRef]
41. Kravchenko, A.; Shevchenko, A.; Ovchinnikov, V.; Grahn, P.; Kaivola, M. Fabrication and characterization of a large-area metal nano-grid wave plate. *Appl. Phys. Lett.* **2013**, *103*, 033111. [CrossRef]

42. Lin, M.Y.; Tsai, T.H.; Kang, Y.L.; Chen, Y.C.; Huang, Y.H.; Chen, Y.J.; Fang, X.; Lin, H.Y.; Choi, W.K.; Wang, L.A.; et al. Design and fabrication of birefringent nano-grating structure for circularly polarized light emission. *Opt. Express* **2014**, *22*, 7388–7398. [CrossRef] [PubMed]
43. Khorasaninejad, M.; Capasso, F. Broadband multifunctional efficient meta-gratings based on dielectric waveguide phase shifters. *Nano Lett.* **2015**, *15*, 6709–6715. [CrossRef] [PubMed]
44. Ra'di, Y.; Sounas, D.L.; Alù, A. Metagratings: Beyond the limits of graded metasurfaces for wave front control. *Phys. Rev. Lett.* **2017**, *119*, 067404. [CrossRef] [PubMed]
45. Mutlu, M.; Akosman, A.E.; Kurt, G.; Gokkavas, M.; Ozbay, E. Experimental realization of a high-contrast grating based broadband quarter-wave plate. *Opt. Express* **2012**, *20*, 27966–27973. [CrossRef] [PubMed]
46. Cong, L.; Cao, W.; Zhang, X.; Tian, Z.; Gu, J.; Singh, R.; Han, J.; Zhang, W. A perfect metamaterial polarization rotator. *Appl. Phys. Lett.* **2013**, *103*, 171107. [CrossRef]

materials

MDPI

Article

Dual-Band Transmissive Cross-Polarization Converter with Extremely High Polarization Conversion Ratio Using Transmitarray

Jianxing Li [1], Jialin Feng [1], Bo Li [2], Hongyu Shi [1], Anxue Zhang [1] and Juan Chen [1,3,4,*]

1 School of Electronic and Information Engineering, Xi'an Jiaotong University, Xi'an 710049, China; jianxingli.china@xjtu.edu.cn (J.L.); fenghui@stu.xjtu.edu.cn (J.F.); honyo.shi1987@gmail.com (H.S.); anxuezhang@xjtu.edu.cn (A.Z.)
2 Institute of Electronic Engineering, China Academy of Engineering Physics, Mianyang 621999, China; libocaepiee@163.com
3 Shenzhen Research School, Xi'an Jiaotong University, Shenzhen 518057, China
4 Guangdong Xi'an Jiaotong University Academy, Foshan 528300, China
* Correspondence: chen.juan.0201@xjtu.edu.cn

Received: 27 April 2019; Accepted: 3 June 2019; Published: 5 June 2019

Abstract: In this paper, a dual-band cross-polarization converter is proposed. The proposed device can convert linearly polarized incident waves to their cross-polarized transmitted waves. Inspired by the aperture coupled transmitarray, a transmissive multi-layered unit cell structure was designed, which can operate in two frequency bands. The designed structure can manipulate the polarization of the transmitted wave into the cross-polarization of the incident waves at 10.36 GHz and 11.62 GHz. The cross-polarized transmittance of the proposed cross-polarization converter is higher than 0.93. In addition, the transmitted wave has an extremely low co-polarized component, which results in a nearly 100% polarization conversion ratio. The two working frequencies can be tuned independently. The proposed cross-polarization converter was simulated, fabricated and measured. The simulation results confirm with the measurement results.

Keywords: cross-polarization converter; transmitarray; high polarization conversion ratio

1. Introduction

Manipulations of microwave properties have attracted great interest due to their applications for wireless communication [1,2], imaging [3–5], stealth technology [6], etc. Techniques using artificial structures, artificial materials or metamaterials have been applied for controlling the polarization [7], amplitude [8], beam shape [9] and direction [10] of microwave. Polarization is one of the basic and important properties of microwaves and has an impact on the performance of communication or radar systems. Polarization converters can be applied for antenna design in microwave communication, remote sensing and imaging systems [11–13]. For instance, by converting a horizontally polarized radar antenna to a vertically polarized radar antenna, a polarization converter can reduce the influences of the ground or sea clutter. In addition, for frequency hopping radar, dual-band and multi-band polarization converter is desired. Polarization of microwaves is usually manipulated by anisotropic structures or artificial materials [14]. Reflective polarization converters are usually thin and wideband, e.g., wideband polarization converters using plasmonic hybridizations [15,16].

The transmissive cross-polarization converters are usually multi-layered artificial structures with complex metallic structures on each layer [17,18]. A single band high efficiency transmissive ultrathin cross-polariztion converter with low in-band co-polarization component was designed using anisotropic artificial structures [19]. Bi-layered chiral metamaterials were also applied for cross-polarization converters from microwave band to terahertz band. However these designs have a co-polarized component of the transmission higher than 0.2 [20–23]. Multi-layered anisotropic metasurface were also used for cross-polarized converter designs, which also has a co-polarized component of the transmitted wave of around 0.2 [24,25]. Multi-layered anisotropic metasurface with metallic gratings can be used for achieving broadband properties. However, the co-polarized component of the transmitted wave can be as high as about 0.1 [26,27]. A dual-band transmissive cross-polarization converter was designed using planar-dipole pair with a co-polarized component of abour 0.15 [28]. Cascaded cavities was used for wideband cross-polarization conversion. However, the co-polarized component is about 0.3 [29].

Reflectarrays or transmitarrays are also used for polarization conversion [30–34]. Reflectarrays were applied for cross-polarization conversion in microwave band and near-infrared band with a co-polarized component of about 0.2 and 0.1, respectively [30,32]. Transmitarrays were applied for single band linear-to-circular polarization conversion [33,34]. Thus, dual-band transmissive cross-polarization converters with extremely low co-polarized component are still in desire.

In this paper, we combined the concepts of transmitarray and polarization converter together and proposed a transmitarray inspired dual-band transmissive cross-polarization converter with extremely low co-polarized component through the whole frequency range. Thus, it can block power transmission in the out-of-band and results in a better frequency selective characteristic. The designed dual-band transmissive cross-polarization converter can convert a linear polarized incident wave to its cross-polarized wave at 10.36 GHz and 11.62 GHz. At these two frequencies, the cross-polarized transmittances are higher than 0.93 and the co-polarized transmittances are suppressed to be 0.0047 and 0.0043, which leads to an almost 100% polarization conversion ratio. In addition, the operating frequencies can be separately tuned by changing the design parameters, which makes this design more useful and general for different practical applications. To our knowledge, it is the first time that the concept of transmitarray is applied for transmissive dual-band cross-polarization converter.

2. Dual-Band Cross-Polarized Converter Design

The proposed cross-polarization converter is a five-layered structure. Such structure is inspired from patch antenna based transmitarray. Different to the previously designed multi layered devices in which all layers contribute to the resonance, only top and bottom layers of the proposed cross-polarization converter resonate. Thus, the insert loss caused by resonances is significantly reduced, which leads to a relatively high efficiency. Structures in each layer of the proposed device unit cell are shown in Figure 1. The blue part and gray part in Figure 1 are substrate and metallic sheet, respectively. The layer-1 and layer-5 are patch elements, which couple and decouple the incident electromagnetic (EM) wave, respectively. These patch elements can be considered as slot coupled square patch antennas. The layer-2 and layer-4 are slots used for wave coupling. Together with layer-2 and layer-4, the layer-3 can be considered as a stripline with a total length s, which contributes to the cross-polarization conversion in the unit cell and can give an additional transmission phase.

As shown in Figure 1a, the structure on layer-1 is a square patch that is fed by a rectangular slot along the x-axis on the layer-2 shown in Figure 1b. The square patch on layer-1 couples the y-polarized incident wave into the unit cell. The incident wave is then coupled to an "L" shaped metallic line on layer-3 through the slot on layer-2. The "L" shaped metallic line is shown in Figure 1c. Such an "L" shaped metallic line transforms the wave propagation along the y-axis into wave propagation along the x-axis. Then, through the rectangular slot along the y-axis on the layer-4 shown in Figure 1d, the waves in the "L" shaped metallic line is coupled to the square patch on layer-5. The layer-5 is the same with the layer-1 as shown in Figure 1e. Layer-5 can decouple the incident wave to x-polarized transmitted wave. Figure 1f shows the side view of the unit cell. Thus, only the layer-1 and layer-5 are resonant structures which help increase the transmission efficiency. The geometric parameters are selected as $a = 11$ mm, $m = 5$ mm, $n = 4.65$ mm, $p = 14$ mm, $v = 4.6$ mm, $l = 8.4$ mm, $w = 0.8$ mm,

$m = 0.7$ mm, $h = 1.5248$ mm, $t = 0.508$ mm and $s = 16$ mm. The dielectric is Taconic TLY-5 with a permittivity of 2.2 and a loss tangent of 0.0009.

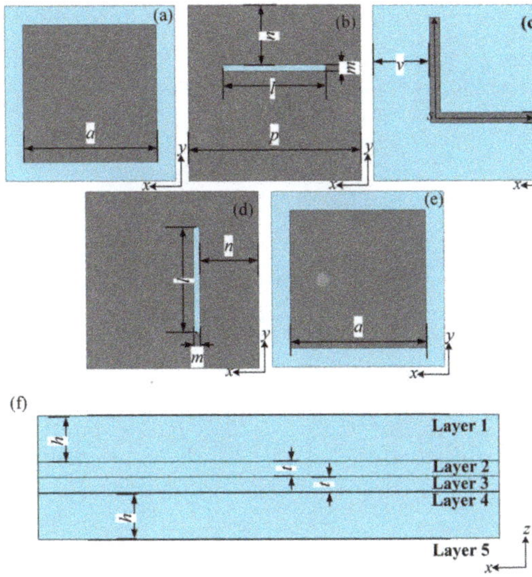

Figure 1. Geometry of the unit cell: (**a**) Layer-1. (**b**) Layer-2. (**c**) Layer-3. (**d**) Layer-4. (**e**) Layer-5. (**f**) Side view.

3. Simulation Results

The unit cell model was built up and simulated in a commercial software CST MICROWAVE STUDIO. In the simulation, the unit cell boundary condition was used along the x- and y-directions, and the absorbing boundary condition was applied for the z-direction. The unit cell model was excited by Flouquet ports with a unit normal incidence of linearly polarized waves along y-axis in the frequency range from 10 GHz to 12 GHz. The amplitude of the transmittances is represented by T. T_{co} is the co-polarized transmittance of the transmitted wave. T_{cr} refers to the cross-polarized transmittance of the transmitted wave. The reflectance is presented by R.

Figure 2a shows the simulation results of T_{cr} and R with an incidence wave propagating along z-axis. The simulation results show that the designed structures can work at 10.36 GHz and 11.62 GHz. At 10.36 GHz, the T_{cr} is 0.93, and the reflectance is 0.126. At 11.62 GHz, the T_{cr} is 0.931, and the reflectance is 0.088. When the T_{cr} reach the maximum, the reflectance is the minimum. The T_{co} is a key parameter of a cross-polarization converter that decides the polarization conversion ratio (PCR). The PCR is defined as $T_{cr}^2/(T_{co}^2 + T_{cr}^2)$ and can directly reflect the polarization purity of the transmitted wave. The simulation result of the T_{co} is shown in Figure 2b. The T_{co} is 0.0047 and 0.0043 at 10.36 GHz and 11.62 GHz, respectively. Thus, according to the numerical data, the PCR was nearly 100% at 10.36 GHz and 11.62 GHz, which demonstrates that this design can obtain an extremely low T_{co} and an extremely high PCR as shown in Figure 3. Thus, the transmitted wave has an extremely high polarization purity.

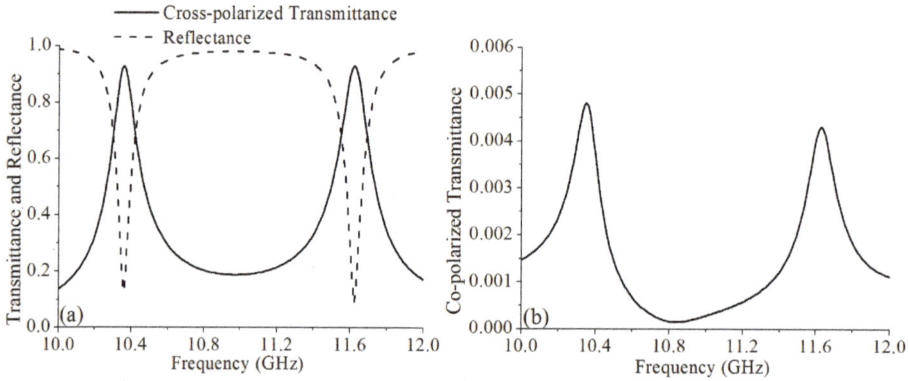

Figure 2. (**a**) The simulated cross-polarized transmittance and reflectance. (**b**) The simulated co-polarized transmittance.

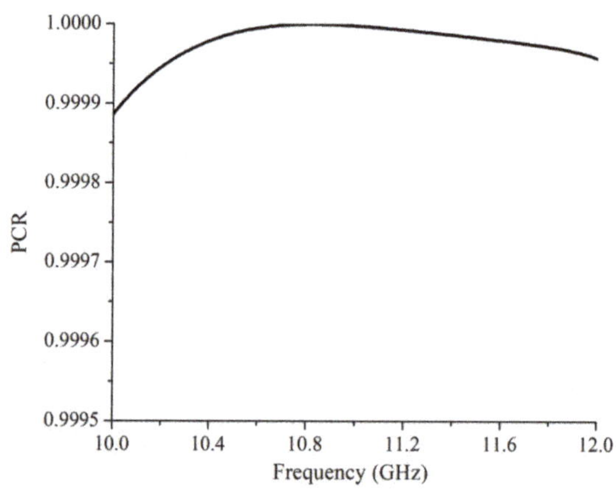

Figure 3. The simulated polarization conversion ratio.

4. Discussion

To provide an insight into the different design parameters, and to illustrate their influence on the frequency behaviour of the proposed cross-polarized converter, parameter sweeps for various a and w were analyzed by simulations. a and w are the key parameters that have significant impacts on the frequency behaviour of the cross-polarized converter. The influences of a are shown in Figures 4 and 5. The influences of w are shown in Figures 6 and 7.

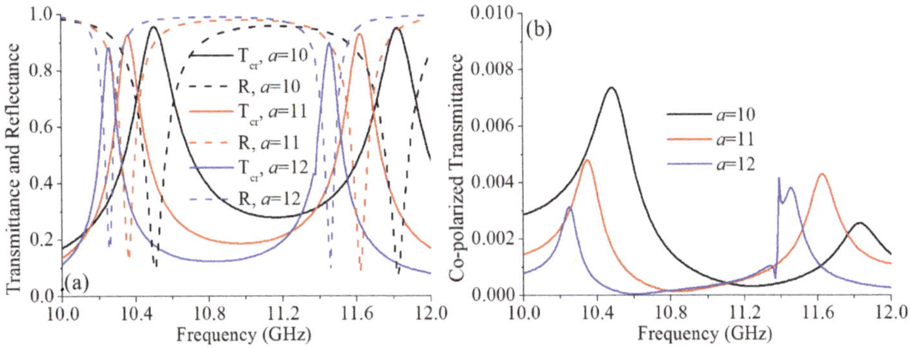

Figure 4. Simulation results with different a: (**a**) The simulated cross-polarized transmittance and reflectance. (**b**) The simulated co-polarized transmittance.

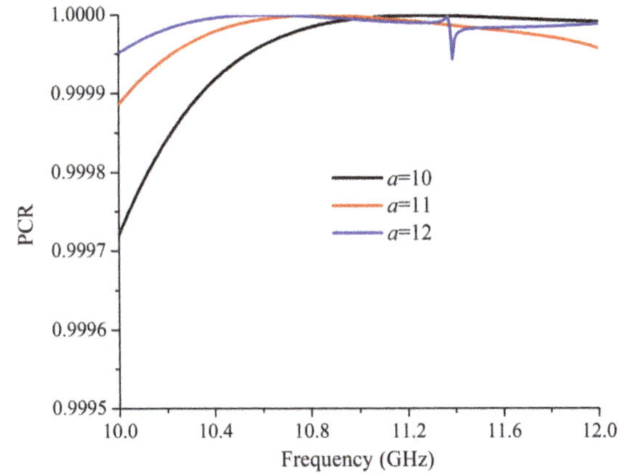

Figure 5. The simulated polarization conversion ratios with different a.

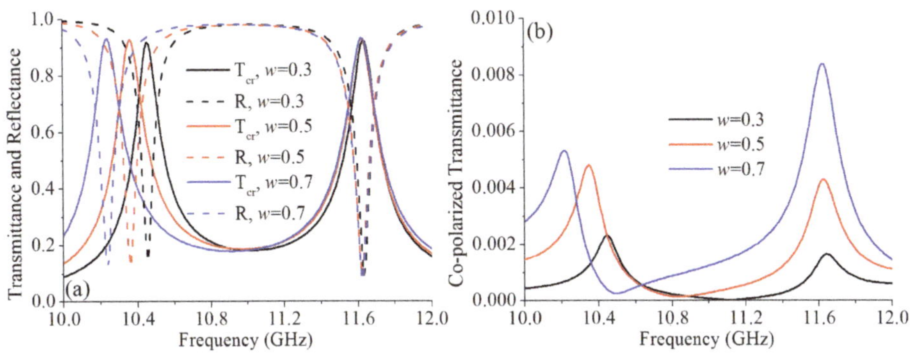

Figure 6. Simulation results with different w: (**a**) The simulated cross-polarized transmittance and reflectance. (**b**) The simulated co-polarized transmittance.

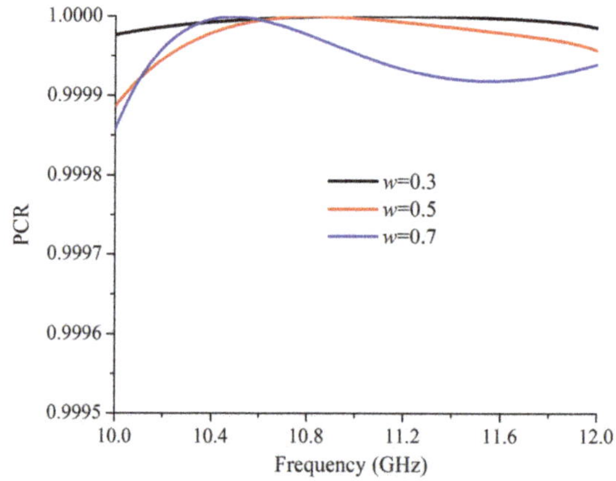

Figure 7. The simulated polarization conversion ratios with different *w*.

As shown in Figure 4a, with a larger *a*, the two operating frequencies had a red shift. When *a* was 10 mm and 12 mm, the working frequency of the cross-polarized converter were 10.5 GHz/11.82 GHz and 10.26 GHz/11.46 GHz, respectively. In addition, with a larger *a*, the cross-polarized transmission peak was higher. The co-polarized transmittances with *a* = 10 mm and *a* = 12 mm are shown in Figure 4b. When *a* = 10 mm, the T_{co} at 10.5 GHz and 11.82 GHz are 0.0073 and 0.0025, respectively. At 10.26 GHz and 11.46 GHz, with *a* = 12 mm, the T_{co} are 0.0031 and 0.0038, respectively. The calculated PCRs with different *a* are shown in Figure 5. When *a* changed, the PCR maintained nearly 100%.

The parameter *w* was another key parameter that impacted on the working frequency. Different to *a*, *w* only influence the lower frequency. As shown in Figure 6a, with a larger *w*, the lower operating frequency has a red shift. When *w* is 0.3 mm and 0.7 mm, the lower working frequency of the cross-polarized converter are 10.45 GHz and 10.24 GHz, respectively. However, when *w* changed, the higher working frequency almost unchanged. Thus, by varying *a* and *w*, the two working frequencies can be tuned independently, which enhances the practicability of the proposed cross-polarization converter. In addition, with a larger *w*, the cross-polarized transmission peak is slightly higher. The co-polarized transmittances with different *w* are shown in Figure 4b. When *w* = 0.3 mm, the T_{co} at 10.45 GHz and 11.62 GHz were 0.0023 and 0.0016, respectively. At 10.24 GHz and 11.62 GHz, with *w* = 0.7 mm, the T_{co} are 0.005 and 0.0084, respectively. The T_{co} was higher with a larger *w*. The calculated PCRs with different *w* are shown in Figure 5. When *w* changed, the PCR maintains nearly 100%.

To further explore the properties of the proposed design under different incident angles, additional simulations were done with incident angles of 10°, 20° and 30°. The simulation results are shown in Figure 8. As shown in Figure 8a, with a larger incident angle within 20°, the lower transmission frequency maintains with a little fluctuation and the higher transmission frequency has a red shift. When the incident angle increased to 30°, the cross-polarization converter is invalidated. The co-polarized component of the transmitted wave maintains extremely low under these incident angles, as shown in Figure 8b. Although the property of the proposed design under oblique incidence is not outstanding, it still can be applied for applications where the polarization converter used with a fixed exciting antenna.

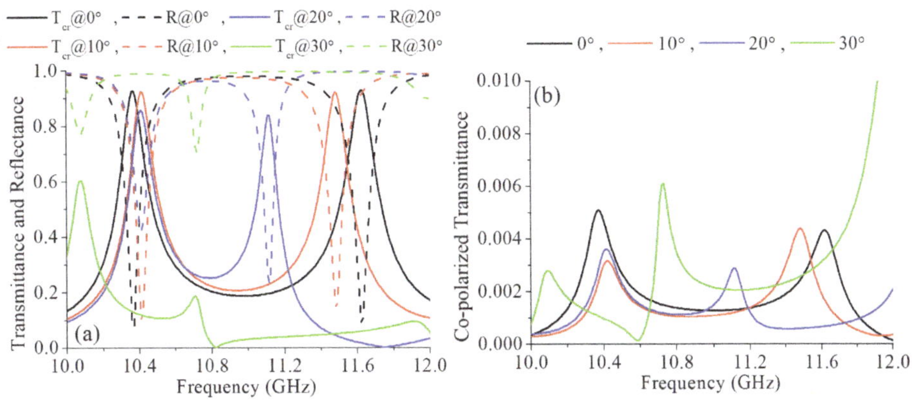

Figure 8. Simulation results with different incident angles: (**a**) The simulated cross-polarized transmittance and reflectance. (**b**) The simulated co-polarized transmittance.

5. Measurement Results

The above proposed design was fabricated by printed circuit board (PCB) processing. The dimension of the sample was 308 mm × 224 mm containing 22 × 16 unit cells. Because the layer-1 and layer-5 are the same, only top view of the fabricated sample is shown in Figure 9.

Figure 9. The fabricated sample of the cross-polarized converter: (**a**) Top view of the sample. (**b**) Side view of the sample.

The measurement setup is shown in Figure 10. Two spot focusing lens horns were used as exciting and receiving antennas. The transmittances and the reflectance were measured by a vector network analyzer (Agilent E8363b). The measured transmittances are shown in Figure 11. The measurement results confirm with the simulated results and their differences are caused by the machining errors, measurement errors and the background noise. The transmittances of the cross-polarization converter sample are shown in Figure 11a. The measured cross-polarized transmittances were 0.935 and 0.914 at 10.37 GHz and 11.71 GHz, respectively. The measured reflectance at these two frequency bands were 0.138 and 0.025, respectively. At 10.37 GHz and 11.71 GHz, the measured co-polarized transmittance was significantly suppressed and was below 0.005 as shown in Figure 11b. The PCR calculated from the measured results is nearly 100% as shown in Figure 12. Thus, the fabricated cross-polarized converter had high efficiency for both transmitted power and polarization conversion.

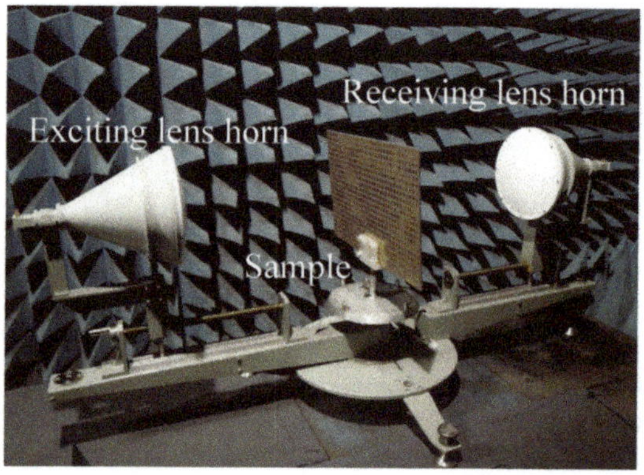

Figure 10. Photo of the measurement setup.

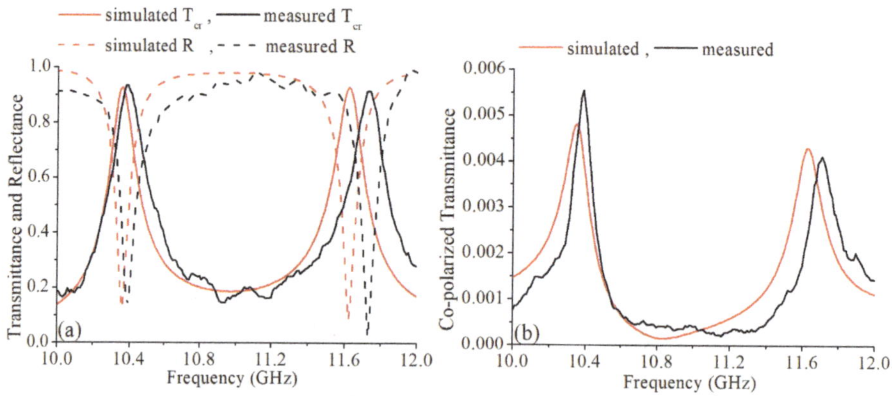

Figure 11. (**a**) The measured and simulated cross-polarized transmittances and reflectances. (**b**) The measured and simulated co-polarized transmittances.

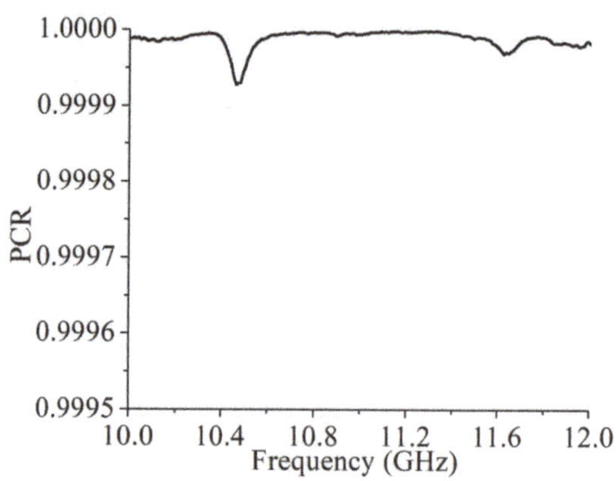

Figure 12. The polarization conversion ratio calculated from the measured results.

6. Conclusions

In conclusion, a dual-band high PCR cross-polarization converter is designed, fabricated and measured. This design is inspired by the concept of transmitarray. The cross-polarized transmittance is higher than 0.9 at 10.37 GHz and 11.71 GHz. The co-polarized transmittance is significantly suppressed below 0.005, which leads to an almost perfect PCR. Thus, the transmitted wave has an extremely high polarization purity. In addition, by varying a and w, the two working frequencies can be tuned independently. The presented design represents a basis for the development of transmissive metasurfaces for wavefront control by changing the length of the stripline on layer-3. The proposed design also can be used for transmissive linear-to-circular polarization converter by using a truncated square patch on the layer-5 instead. The proposed cross-polarization converter can be employed in antenna, radar or telecommunication applications.

Author Contributions: Conceptualization, S.H.; methodology, J.L. and F.J.; validation, J.L.; formal analysis, J.L. and F.J.; investigation, F.J. and H.S.; resources, A.Z.; data curation, B.L.; writing—original draft preparation, H.S.; writing—review and editing, H.S.; visualization, J.L. and F.J.; supervision, J.L. and J.C.; project administration, J.C., A.Z. and H.S.; funding acquisition, J.L.

Funding: This research was funded by National Natural Science Foundation of China grant number 61871315, in part by Technology Program of Shenzhen grant number JCYJ20170816100722642, JCYJ20180508152233431 and in part by the Natural Science Foundation of Guangdong Province, China grant number 2018A030313429. The APC was funded by 61871315.

Conflicts of Interest: The authors declare no conflict of interest.

References

1. Gupta, S.; Briqech, Z.; Sebak, A.R.; Denidni, T.A. Mutual-Coupling Reduction Using Metasurface Corrugations for 28 GHz MIMO Applications. *IEEE Antennas Wirel. Propag. Lett.* **2017**, *16*, 2763–2766. [CrossRef]
2. Johnson, M.C.; Brunton, S.L.; Kundtz, N.B.; Kutz, J.N. Sidelobe Canceling for Reconfigurable Holographic Metamaterial Antenna. *IEEE Trans. Antennas Propag.* **2015**, *63*, 1881–1886. [CrossRef]
3. Fan, Q.; Xu, T. Research progress of imaging technologies based on electromagnetic metasurfaces. *Acta Phys. Sin.* **2017**, *66*. [CrossRef]
4. Liu, K.; Cheng, Y.Q.; Yang, Z.C.; Wang, H.Q.; Qin, Y.L.; Li, X. Orbital-Angular-Momentum-Based Electromagnetic Vortex Imaging. *IEEE Antennas Wirel. Propag. Lett.* **2015**, *14*, 711–714. [CrossRef]
5. Zhao, M.; Zhu, S.; Chen, X.; Li, J.; Hu, D.; Wang, L.; Zhang, A. Frequency-Diverse transmission metamaterial aperture with a bunching random beam. *IEEE Antennas Wirel. Propag. Lett.* **2018**, *17*, 1029–1033. [CrossRef]
6. Schurig, D.; Mock, J.J.; Justice, B.J.; Cummer, S.A.; Pendry, J.B.; Starr, A.F.; Smith, D.R. Metamaterial electromagnetic cloak at microwave frequencies. *Science* **2006**, *314*, 977–980. [CrossRef]
7. Shi, H.Y.; Zhang, A.X.; Zheng, S.; Li, J.X.; Jiang, Y.S. Dual-band polarization angle independent 90 degrees polarization rotator using twisted electric-field-coupled resonators. *Appl. Phys. Lett.* **2014**, *104*. [CrossRef]
8. Li, A.; Kim, S.; Luo, Y.; Li, Y.; Long, J.; Sievenpiper, D.F. High-Power Transistor-Based Tunable and Switchable Metasurface Absorber. IEEE Trans. *Microw. Theory Tech.* **2017**, *65*, 2810–2818. [CrossRef]
9. Chen, M.L.L.N.; Jiang, L.J.; Sha, W.E.I. Ultrathin Complementary Metasurface for Orbital Angular Momentum Generation at Microwave Frequencies. *IEEE Trans. Antennas Propag.* **2017**, *65*, 396–400. [CrossRef]
10. Shi, H.; Li, J.; Zhang, A.; Jiang, Y.; Wang, J.; Xu, Z.; Xia, S. Gradient Metasurface With Both Polarization-Controlled Directional Surface Wave Coupling and Anomalous Reflection. *IEEE Antennas Wirel. Propag. Lett.* **2015**, *14*, 104–107. [CrossRef]
11. Suss, H.; Gruner, K.; Wilson, W.J. Passive Millimeter-Wave Imaging a Tool for Remote-Sensing. *Alta Freq.* **1989**, *58*, 457–465.
12. Euler, M.; Fusco, V.; Cahill, R.; Dickie, R. 325 GHz Single Layer Sub-Millimeter Wave FSS Based Split Slot Ring Linear to Circular Polarization Convertor. *IEEE Trans. Antennas Propag.* **2010**, *58*, 2457–2459. [CrossRef]
13. Dietlein, C.; Luukanen, A.; Popovic, Z.; Grossman, E. A W-band polarization converter and isolator. *IEEE Trans. Antennas Propag.* **2007**, *55*, 1804–1809. [CrossRef]

14. Liu, F.; Liang, Z.X.; Li, J.S. Manipulating Polarization and Impedance Signature: A Reciprocal Field Transformation Approach. *Phys. Rev. Lett.* **2013**, *111*. [CrossRef] [PubMed]
15. Shi, H.; Li, J.; Zhang, A.; Wang, J.; Xu, Z. Broadband cross polarization converter using plasmon hybridizations in a ring/disk cavity. *Opt. Express* **2014**, *22*, 20973–20981. [CrossRef] [PubMed]
16. Dong, G.; Shi, H.; Xia, S.; Zhang, A.; Xu, Z.; Wei, X. Ultra-broadband perfect cross polarization conversion metasurface. *Opt. Commun.* **2016**, *365*, 108–112. [CrossRef]
17. Pfeiffer, C.; Grbic, A. Metamaterial Huygens' Surfaces: Tailoring Wave Fronts with Reflectionless Sheets. *Phys. Rev. Lett.* **2013**, *110*, 197401. [CrossRef] [PubMed]
18. Achouri, K.; Khan, B.A.; Gupta, S.; Lavigne, G.; Salem, M.A.; Caloz, C. Synthesis of electromagnetic metasurfaces: principles and illustrations. *EPJ Appl. Metamater.* **2016**, *2*. [CrossRef]
19. Xu, P.; Jiang, W.X.; Wang, S.Y.; Cui, T.J. An Ultrathin Cross-Polarization Converter with Near Unity Efficiency for Transmitted Waves. *IEEE Trans. Antennas Propag.* **2018**, *66*, 4370–4373. [CrossRef]
20. Wu, S.; Xu, S.; Zinenko, T.L.; Yachin, V.V.; Prosvirnin, R.L.; Tuz, V.R. 3D-printed chiral metasurface as a dichroic dual-band polarization converter. *Opt. Lett.* **2019**, *44*, 1056–1059. [CrossRef]
21. Huang, C.; Feng, Y.J.; Zhao, J.M.; Wang, Z.B.; Jiang, T. Asymmetric electromagnetic wave transmission of linear polarization via polarization conversion through chiral metamaterial structures. *Phys. Rev. B* **2012**, *85*. [CrossRef]
22. Zhao, J.; Cheng, Y. Ultrathin dual-band polarization angle independent 90 degrees polarization rotator with giant optical activity based on planar chiral metamaterial. *Appl. Phys. B-Lasers Opt.* **2018**, *124*. [CrossRef]
23. Cheng, Z.; Cheng, Y. A multi-functional polarization convertor based on chiral metamaterial for terahertz waves. *Opt. Commun.* **2019**, *435*, 178–182. [CrossRef]
24. Li, W.; Xia, S.; Shi, H.; Li, J.; Zhang, A.; Li, Z.; Xu, Z. Design of a Dual-Band Dual-Polarization Transparent Frequency Selective Surface. *IEEE Antennas Wirel. Propag. Lett.* **2017**, *16*, 3172–3175. [CrossRef]
25. Zhang, L.; Zhou, P.; Chen, H.; Lu, H.; Xie, H.; Zhang, L.; Li, E.; Xie, J.; Deng, L. Ultrabroadband Design for Linear Polarization Conversion and Asymmetric Transmission Crossing X- and K- Band. *Sci. Rep.* **2016**, *6*. [CrossRef] [PubMed]
26. Zhou, G.; Tao, X.; Shen, Z.; Zhu, G.; Jin, B.; Kang, L.; Xu, W.; Chen, J.; Wu, P. Designing perfect linear polarization converters using perfect electric and magnetic conducting surfaces. *Sci. Rep.* **2016**, *6*. [CrossRef] [PubMed]
27. Song, K.; Liu, Y.; Luo, C.; Zhao, X. High-efficiency broadband and multiband cross-polarization conversion using chiral metamaterial. *J. Phys. D-Appl. Phys.* **2014**, *47*. [CrossRef]
28. Wang, S.; Liu, W.; Wen, G. Dual-band transmission polarization converter based on planar-dipole pair frequency selective surface. *Sci. Rep.* **2018**, *8*. [CrossRef] [PubMed]
29. Wang, J.; Shen, Z.; Wu, W. Cavity-based high-efficiency and wideband 90 degrees polarization rotator. *Appl. Phys. Lett.* **2016**, *109*. [CrossRef]
30. Deguchi, H.; Higashi, D.; Yamada, H.; Matsumoto, S.; Tsuji, M. Arbitrarily-Shaped Reflectarray Resonant Elements for Dual-Polarization Use and Polarization Conversion. *IEICE Trans. Commun.* **2018**, *E101B*, 277–284. [CrossRef]
31. Hung, L.; Yang, Z.; Gao, M.; Booske, J.H.; Behdad, N. A Wideband, Single-Layer Reflectarray Exploiting a Polarization Rotating Unit Cell. *IEEE Trans. Antennas Propag.* **2019**, *67*, 872–883. [CrossRef]
32. Wang, R.; Li, L.; Tian, H.; Liu, J.; Liu, J.; Tian, F.; Zhang, J.; Sun, W. Full telecomband covered half-wave meta-reflectarray for efficient circular polarization conversion. *Opt. Commun.* **2018**, *427*, 469–476. [CrossRef]
33. Pfeiffer, C.; Grbic, A. Millimeter-Wave Transmitarrays for Wavefront and Polarization Control. *IEEE Trans. Microw. Theory Tech.* **2013**, *61*, 4407–4417. [CrossRef]
34. Huang, C.; Pan, W.; Ma, X.; Zhao, B.; Cui, J.; Luo, X. Using Reconfigurable Transmitarray to Achieve Beam-Steering and Polarization Manipulation Applications. *IEEE Trans. Antennas Propag.* **2015**, *63*, 4801–4810. [CrossRef]

materials

MDPI

Article

Active Enhancement of Slow Light Based on Plasmon-Induced Transparency with Gain Materials

Zhaojian Zhang [1], Junbo Yang [2,*], Xin He [2], Yunxin Han [2], Jingjing Zhang [1], Jie Huang [1], Dingbo Chen [1] and Siyu Xu [1]

1 College of Liberal Arts and Sciences, National University of Defense Technology, Changsha 410073, China; 376824388@sjtu.edu.cn (Z.Z.); zhangjingjing13@nudt.edu.cn (J.Z.); jhuang_nudt@163.com (J.H.); c_dingbo@163.com (D.C.); 13149600390@163.com (S.X.)

2 Center of Material Science, National University of Defense Technology, Changsha 410073, China; xinhestudy@163.com (X.H.); hanyx15@163.com (Y.H.)

* Correspondence: yangjunbo@nudt.edu.cn

Received: 8 May 2018; Accepted: 31 May 2018; Published: 3 June 2018

Abstract: As a plasmonic analogue of electromagnetically induced transparency (EIT), plasmon-induced transparency (PIT) has drawn more attention due to its potential of realizing on-chip sensing, slow light and nonlinear effect enhancement. However, the performance of a plasmonic system is always limited by the metal ohmic loss. Here, we numerically report a PIT system with gain materials based on plasmonic metal-insulator-metal waveguide. The corresponding phenomenon can be theoretically analyzed by coupled mode theory (CMT). After filling gain material into a disk cavity, the system intrinsic loss can be compensated by external pump beam, and the PIT can be greatly fueled to achieve a dramatic enhancement of slow light performance. Finally, a double-channel enhanced slow light is introduced by adding a second gain disk cavity. This work paves way for a potential new high-performance slow light device, which can have significant applications for high-compact plasmonic circuits and optical communication.

Keywords: plasmon-induced transparency; metal-dielectric-metal; gain material

1. Introduction

Electromagnetically induced transparency (EIT), which arises from the quantum destructive interference between two distinct excitation channels in a three-level atomic system [1], has potential in slow light [2] and nonlinear optical response enhancement [3] due to its excellent ability to modulate dispersion. However, EIT requires strict experimental conditions such as low temperature and stable pumping [1], which makes it hard to be applied in practical optical systems, especially on-chip devices. As an analogue of EIT in a plasmonic system, plasmon-induced transparency (PIT) has drawn more attention. PIT is attributed to the destructive interference between plasmonic radiative mode and subradiative mode [4], which can be realized on metamaterials [4] and nanoscale plasmonic circuits [5]. Nowadays, PIT can be found extensively in nano applications for sensing [6–8], filtering [9,10], and slow light [11–16].

Although PIT can be tuned by changing geometric parameters, it is essential to propose an active modulation of PIT in practical applications, and thus various optical materials are utilized to achieve this target. Superconductor and polymer are used to obtain temperature modulation of transparent window position [9,17], and nonlinear Kerr material can realize ultrafast all-optical control [13,18]. Utilizing 2D material graphene, both transparent window position and intensity can be dynamically adjusted by external electric gating voltage [19–21]. Although tunable PIT ensures good control of delay time for slow light applications, such delay time always has a threshold because the dispersion modulation of PIT is restricted by the intrinsic loss (ohmic loss) of metal. In particular, the delay time can only reach approximately 1 picosecond (ps) at most [14] in the plasmonic metal-dielectric-metal (MDM) system. Some studies have found that PIT can be dramatically enhanced by applying gain material to overcome the loss, and such gain-assisted plasmonic devices exhibit brilliant performance as a spaser (nanolaser) [22,23], sensor [24,25], and buffer [26,27]. However, few works have focused on gain-assisted PIT in MDM system. Compared to dielectric waveguide, plasmonic MDM waveguide can confine optical modes in deep subwavelength to break the diffraction limit, which is a solution for the next generation of highly integrated on-chip circuits [28]. Pumped active plasmonic waveguides may suffer from the self-heating effect [29], but the problem can be solved by a heat sink system [30].

In this paper, PIT is investigated numerically based on plasmonic MDM waveguide system side-coupled with a stub and disk resonator. The corresponding phenomenon can be theoretically analyzed by coupled mode theory (CMT). After filling gain material into a disk cavity, PIT can be greatly boosted by external pump beam to achieve a dramatic enhancement of slow light performance. Finally, a double-channel enhanced slow light is introduced by adding a second gain disk cavity. This work paves way for a potential new high-performance slow light device, which can have significant applications in high-compact plasmonic circuits and optical communication.

2. Materials and Methods

In Figure 1a, the plasmonic MDM waveguide system is presented in 3D image, and the 2D scheme from the view of *z*-axis is shown in Figure 1b; the detailed structural geometric parameters are given in the caption. Here, nickel (Ni) is utilized to promote adhesion between metal and substrate [31]. For the height of the MDM waveguide, the relationship between the surface plasmon waves (SPWs) effective refractive index (n_{eff}) and the height at the wavelength of 1310 nm (which is the main wavelength discussed in this paper) is given in Figure 1c, where the width of waveguide is fixed at 100 nm. When the height rises at start, both real and imaginary part of the n_{eff} will drop owing to the reduction of the modal power fraction at interfaces [32]. The decrease in $Im\left(n_{eff}\right)$ indicates less propagation loss and, consequently, longer propagation length. However, after reaching a certain height, n_{eff} will keep stable. Therefore, the height can be selected as 100 nm according to the outcome. The distribution $|P_x|$ of the fundamental mode at 1310 nm when height is 100 nm is shown in Figure 1d.

To describe the permittivity of Ag, the Drude model is utilized as follows [6]:

$$\varepsilon_{\mathrm{m}} = \varepsilon_{\infty} - \frac{\omega_p^2}{\omega(\omega + i\gamma)} \tag{1}$$

where ε_{∞} is the dielectric permittivity of the infinite frequency, ω_p refers to the bulk frequency for plasma, γ is the damping frequency for electron oscillation, and ω gives the incident light angular frequency. The corresponding parameters of Ag are ε_{∞} = 3.7, ω_p = 1.38 × 10^{16} Hz, and γ = 2.73 × 10^{13} Hz. The stub and disk cavities are both filled with silicon (Si, ε = 12.25) [33]. To ensure the stability of the guided light propagation as well as resist the corrosion and oxidation, the waveguide should be filled with dielectric materials. For optical interconnects, optical polymer is a good choice for several reasons [34]. First, polymer has the compatibility with existing manufacturing processes of conventional plasmonic waveguides. Second, it can be easily integrated into existing architectures by the spin coating method. Third—and most importantly—it can withstand the high-temperature environment of a pumped system. Additionally, it is low-cost. Here, sodium p-styrenesulfonate (PSSNa) homopolymer with a refractive index of 1.395 [31] is selected. Compared with other conventional optical polymers such as polymethyl methacrylate (PMMA) [35], PSSNa has a lower refractive index and therefore suffers from less propagation loss.

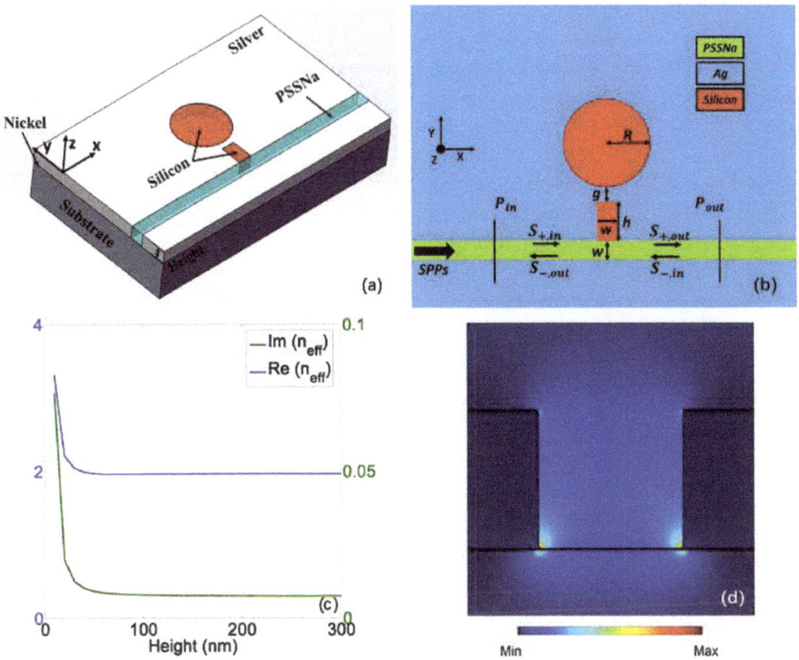

Figure 1. (**a**) 3D scheme of the plasmonic metal-dielectric-metal (MDM) waveguide system; (**b**) 2D scheme of this system from the view of *z*-axis. The geometry parameters are w = 100 nm, h = 205 nm, g = 19 nm, R = 141 nm; (**c**) The relationship between the effective refractive index (n_{eef}) of surface plasmon waves (SPWs) and the height at a wavelength of 1310 nm; (**d**) The distribution $|P_x|$ of the fundamental mode at 1310 nm when height is 100 nm.

In MDM waveguide, only transverse-magnetic (TM) mode can exist [36]. Compared to incident wavelength, the width of the waveguide is much smaller, so there is only fundamental TM mode. The dispersion relation of this fundamental mode is described as follows [36]:

$$\frac{\varepsilon_i p}{\varepsilon_m k} = \frac{1 - e^{kw}}{1 + e^{kw}}$$

$$k = k_0\sqrt{\left(\frac{\beta_{spp}}{k_0}\right)^2 - \varepsilon_i},\ p = k_0\sqrt{\left(\frac{\beta_{spp}}{k_0}\right)^2 - \varepsilon_m} \tag{2}$$

$$\beta_{spp} = n_{eff}k_0 = n_{eff}\frac{2\pi}{\lambda}$$

Here, w refers to the width of the waveguide, λ is incident light wavelength in vacuum, ε_i and ε_m give the dielectric and metal permittivity, β_{spp} is propagation constant of SPWs, and $k_0 = 2\pi/\lambda$ is the wave number. The 2D finite-difference time-domain (FDTD) solution with mesh grid size 2 nm is utilized to simulate this device with the boundary condition of stabilized perfectly matched layers (PML) to maintain convergence. To collect the incident and transmitted power, two monitors are put at P_{in} and P_{out}, respectively, as shown in Figure 1b. The transmission spectrum of power is calculated as $T = P_{out}/P_{in}$.

The transmission spectrum of this structure from FDTD is depicted in Figure 2a, which possesses a PIT profile. The spectrum without disk cavity is given in the inset, showing a band-stop spectral characteristic. The corresponding distribution of H_z are given in Figure 2b–e. In a plasmonic waveguide system, PIT results from the destructive interference between radiative mode (directly excited mode) and subradiative mode (indirectly excited mode). Here, both stub and disk cavities can act as resonators. The stub cavity can be seen as a Fabry–Perot (F–P) resonator, with the F–P mode (FPM) corresponding to the radiative mode. The disk resonator possesses a whispering gallery mode (WGM) that serves as the subradiative mode. The resonance conditions are respectively given as follows [37]:

$$\begin{array}{c} FPM : m\lambda = 2h \cdot \mathrm{Re}(n_{eff}), m = 1,2\ldots \\ WGM : k_d \frac{H_n^{(1)'}(k_e R)}{H_n^{(1)}(k_e R)} = k_e \frac{J_n'(k_d R)}{J_n(k_d R)}, n = 1,2\ldots \end{array} \tag{3}$$

where R is the disk radius, k_d and k_e are the wave vectors in the disk resonator and metal, respectively, $H_n^{(1)}$ and $H_n^{(1)'}$ are the first kind Hankle function with order n and its derivative, respectively, and J_n and J_n' are the first kind Bessel function with order n and its derivative, respectively. m and n refer to the mode number which is an integer. According to the field distribution in Figure 2, $m = 1$ and $n = 2$. The generating mechanism of PIT can be theoretically analyzed by CMT [15]. As demonstrated in Figure 1b, $S_{\pm,in}$ and $S_{\pm,out}$ represent amplitudes of input and output SPWs, respectively; the subscript $\pm$ means two directions of wave propagation. The decay rates of stub and disk resonators are indicated as α and β, respectively, which arises from the resonator intrinsic loss. The coupling coefficient between the bus waveguide and stub resonator is given as γ, and δ is the coupling coefficient between the two resonators. According to temporal CMT, the field amplitudes a and b that correspond to FPM and WGM, respectively, can be described as follows [15]:

$$\begin{array}{l} \frac{da}{dt} = (j\omega_0 - \alpha - \gamma)a + j\sqrt{\gamma}(S_{+in} + S_{-in}) + j\delta b \\ \frac{db}{dt} = (j\omega_0 - \beta)b + j\delta a \end{array} \tag{4}$$

where ω_0 is the common resonant frequency of two resonators, which is also the central frequency of the transparent window, and j is an imaginary unit. According to power conservation and time reversal symmetry, the input and output wave amplitudes have a relationship as follows:

$$\begin{array}{l} S_{+,out} = S_{+,in} + j\sqrt{\gamma}a \\ S_{-,out} = S_{-,in} + j\sqrt{\gamma}a \end{array} \tag{5}$$

Here, $S_{-,in} = 0$ because SPWs is input from the left. Based on Equations (4) and (5), the transmission spectrum can be calculated as:

$$T(\omega) = \left|\frac{S_{+,out}}{S_{+,in}}\right|^2 = \left|1 - 2\gamma \frac{2j(\omega - \omega_0) + 2\beta}{[2j(\omega - \omega_0) + \alpha + \beta + \gamma]^2 + (2\delta)^2 - (\alpha - \beta + \gamma)^2}\right|^2 \tag{6}$$

The transmission data from CMT is also plotted in Figure 2a, showing that the theoretical curve is in great agreement with the simulated data.

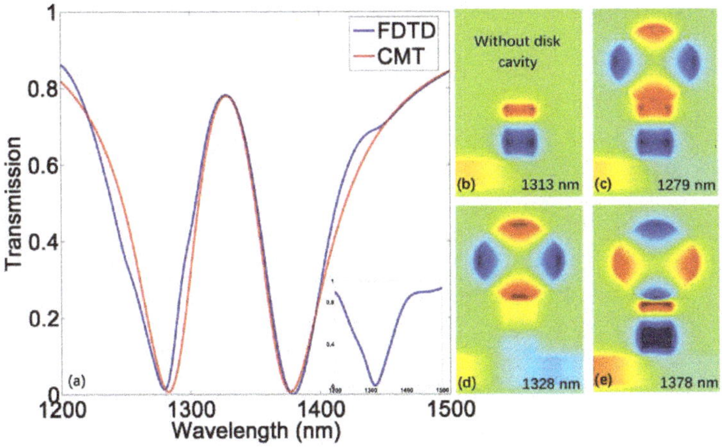

Figure 2. (**a**) The transmission spectrum from finite-difference time-domain (FDTD) and coupled mode theory (CMT). The inset is the transmission spectrum without disk cavity; (**b**–**e**) The corresponding H_z distribution. For plasmon-induced transparency (PIT), the central transparent wavelength is 1328 nm, the valley values are at 1279 nm and 1378 nm.

3. The Slow Light Performance

Slow light effect is an indispensable part of PIT applications. The spectrum profile of PIT indicates that there is an extreme modification of the dispersion properties within the transparent window. In an anomalous dispersion regime, a group velocity large than c (the speed of light in vacuum) leads to a superluminal pulse propagation, and such fast light can be utilized for gravitational wave detection and rotation sensing [38]. In a normal dispersion regime, the group velocity can be slower than c and a subluminal pulse propagation can be obtained, which is called slow light [39]. Here we only focus on the slow light effect. The slow light performance can be assessed by optical delay time τ_g and group index n_g described as follows [12]:

$$\begin{array}{l} \tau_g = \frac{d\psi(\omega)}{d\omega} \\ n_g = \frac{c}{v_g} = \frac{c}{D}\tau_g \end{array} \quad (7)$$

where $\psi(\omega)$ stands for the transmission phase shift from the light source to the monitor, c is the light speed, v_g is the group velocity in the plasmonic waveguide, and D is the length of this system. Since the length of the different devices are not the same, it is more accurate to evaluate the slow light performance by optical delay time. Figure 3a,b show the phase shift and delay time of the device in Figure 1b; the positive and negative delay time represent slow and fast light effect, respectively. We can find a slope of phase shift within the transparent window, producing a delay time of approximately 0.074 ps around the PIT peak. Such values are low because this PIT profile is not very sharp, consequently producing a relatively gentle dispersion. It is known that a wider gap between two resonators can reduce the corresponding coupling coefficient to produce a narrower PIT peak with a steeper slope [14,15], which can lead to a stronger dispersion. However, it will also bring more intrinsic loss of metal, causing lower transmission of slow light that will make the device inefficient.

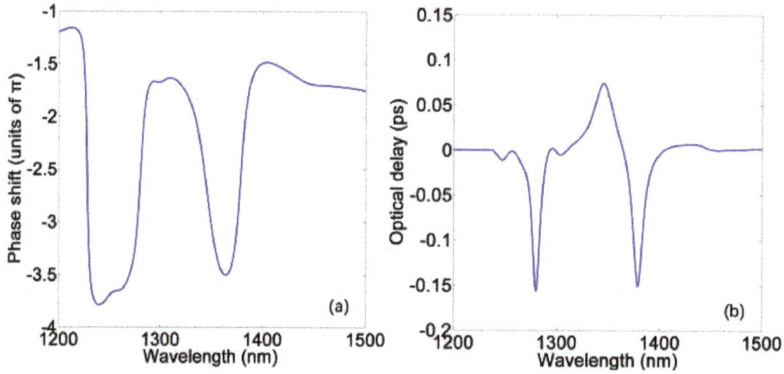

Figure 3. (**a**) The spectrum of phase shift in no-gain PIT system; (**b**) The spectrum of delay time in no-gain PIT system.

4. Slow Light Enhanced by Gain Material

To compensate for the intrinsic loss in the plasmonic system, optical gain materials such as quantum dots and dye molecules have been utilized [40,41]. Here, we use the semiconductor InGaAsP ($\varepsilon = 11.38 - i\varepsilon_I$) [33] as an active media to fill the disk cavity as shown in Figure 4a. The interaction between pump beam and gain medium can be described by a four-level quantum system shown in Figure 4b [24]. Inside the gain media, the pump photons can excite an electronic transition from the ground state $|0\rangle$ to the highest excited state $|3\rangle$, then come to a metastable state $|2\rangle$ via a fast nonradiative transition. Next, the signal light will act as the trigger to make gain material return to a lower level $|1\rangle$ by radiating photons with same frequency as signal light. Consequently, pump energy is transferred to the plasmonic system, providing the compensation for the system intrinsic loss. One of the schemes to introduce pump beam is from the input of waveguide. However, the most common pump wavelength for InGaAsP is 980 nm [42,43], which cannot be coupled into gain material as there is no resonant behavior inside such disk cavity at 980 nm. Therefore, pump beam should be directly injected on the gain material from the top (*z*-axis) of this device [44].

In this simulation, the imaginary part of the active medium permittivity can represent the loss or gain. ε_I will be -0.1 without the pump beam, and the positive imaginary part of permittivity indicates the case of loss. When increasing the pump power, ε_I will rise and eventually become positive, and the medium will show the gain effect with a negative imaginary part of dielectric constant. Here, the gain coefficient $\eta = -(2\pi/\lambda)Im\sqrt{11.38 - i\varepsilon_I}$ is utilized to describe the gain level [22].

Figure 4c,d demonstrate the PIT transmission spectrum with different gain coefficients of which central transparent wavelength is at conventional telecommunication wavelength of 1310 nm. When $\eta = -710$ cm^{-1} ($\varepsilon_I = -0.1$), which corresponds to no pumping [33], there is no transparent peak. Because the system is lossy under this condition, the electric field in the disk attenuates quickly so the WGM (subradiative mode) cannot be formed. As the pump power increases, the intrinsic loss of system will be compensated gradually. Therefore, the suppressed PIT will be released, leading to the rise of a transparent peak. When the peak value reaches the unity, it indicates that the metallic intrinsic loss is fully compensated by the gain. After that, the surplus gain power will be efficiently delivered to the PIT resonance via energy matching as the gain level continues increasing, leading to the amplification of the SPWs. Therefore, the transparent peak will exceed the unity and keep rising as shown in Figure 4d. [24]. Such phenomenon can be applied to realize nanolasers [23].

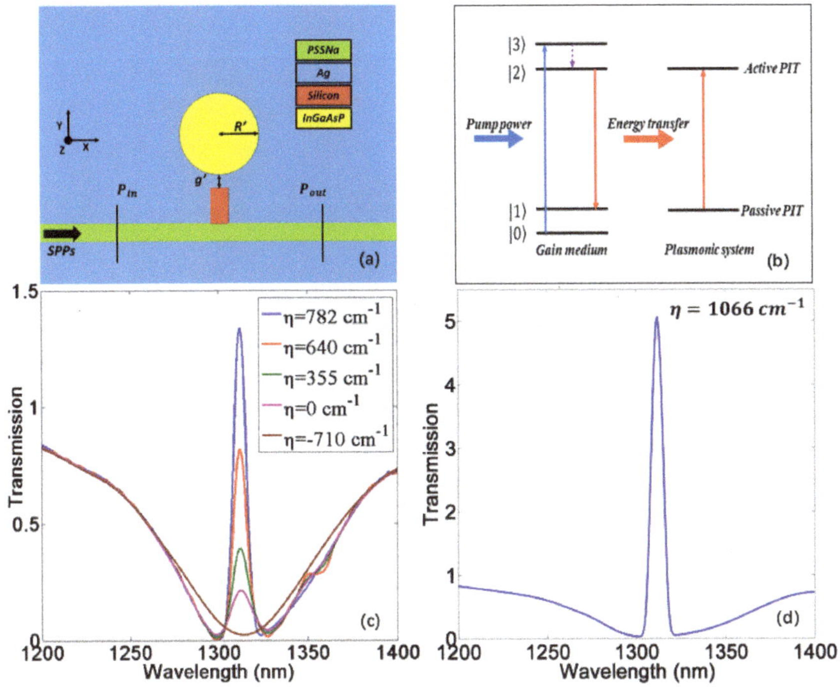

Figure 4. (**a**) 2D scheme of gain-assisted PIT system from the view of z-axis. The geometry parameters are $g' = 41$ nm, $R' = 149$ nm; (**b**) Schematic of the energy transfer from a pumped four-level gain medium to the PIT resonance in plasmonic system; (**c**,**d**) The transmission spectrum with different gain coefficients.

For a slow light system, however, the excessive gain may risk damage to the detector due to the ultrahigh output intensity. Therefore, we only focus on the slow light effect of PIT with peak value near unity. The phase shift and delay time corresponding to three different gain levels are exhibited in Figure 5a–f. These show that as the gain level rises, the transmission peak will increase and the delay time is dramatically enhanced simultaneously. Figure 6 displays the relationship between transmission/delay time and gain coefficient at 1310 nm. When $\eta = 782$ cm^{-1}, the delay time can be improved up to 2.4 ps in this plasmonic MDM system with a high transmission at 1310 nm, which is greater than any previous similar passive systems [12–16]. At the same time, the transmission of transparent peak is 1.34, which is only a little higher than the unity. Such excellent slow light performance can be attributed to two reasons. First, the gain compensation for the ohmic loss allows for a wider gap between two resonators and consequently produces a narrow transparent peak with considerable amplitude. Second, the high gain level can boost the peak value, making the peak slope steeper. Such extremely sharp PIT profile will lead to a strong group velocity dispersion at the transparent range. It has to be mentioned that the highest gain level above is still within the range of current semiconductor gain materials [45,46].

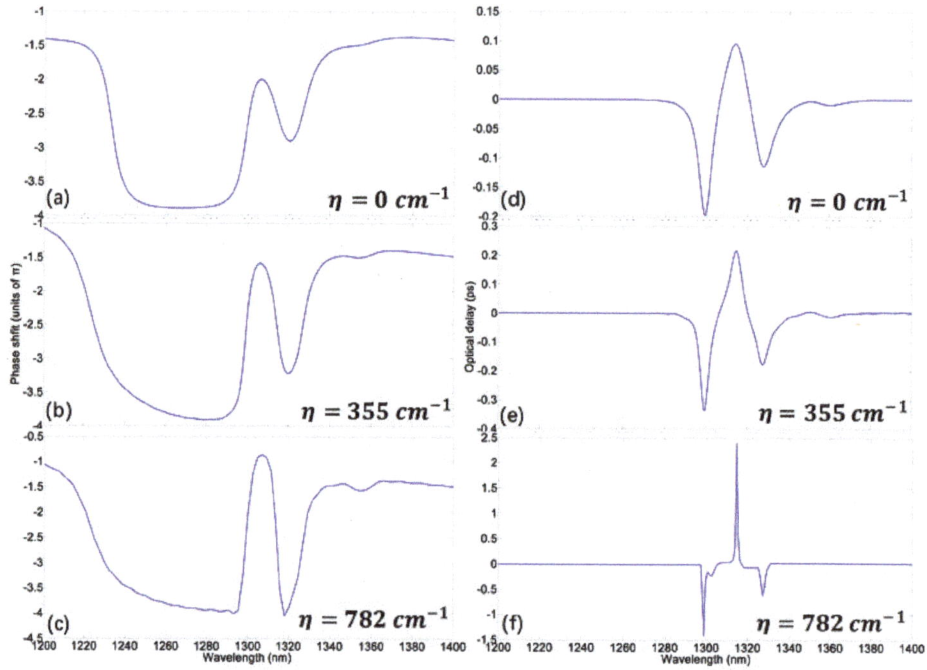

Figure 5. (a–f) The phase shift and delay time corresponding to three different gain levels.

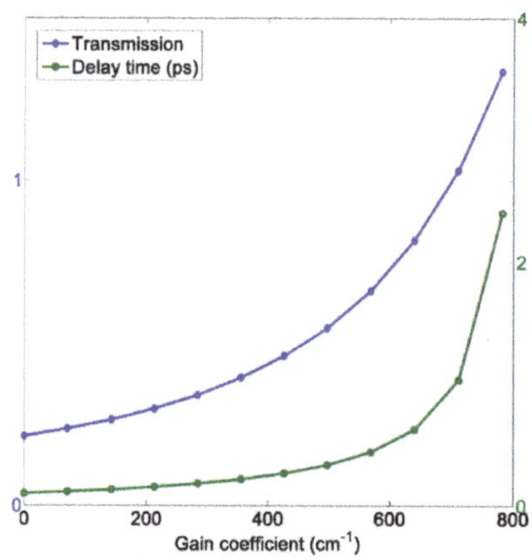

Figure 6. The relationship between transmission/delay time and gain coefficients at 1310 nm.

5. Gain-Assisted Slow Light with Double Channels

By introducing the second gain-filled disk cavity, double boosted PIT can be achieved as shown in Figure 7a,b. The field distributions corresponding to the two transparent windows are given in the inset of Figure 7b. From this, we can see that the combination of different WGMs can produce two kinds of subradiative modes, leading to the double PIT. Under the gain level η = 7820 cm^{-1}, both transparent peaks can be fueled by the active gain, bringing double-channel enhanced slow light performance as shown in Figure 7c,d. At two different channels, the delay time can reach 1.37 ps and 0.59 ps, respectively.

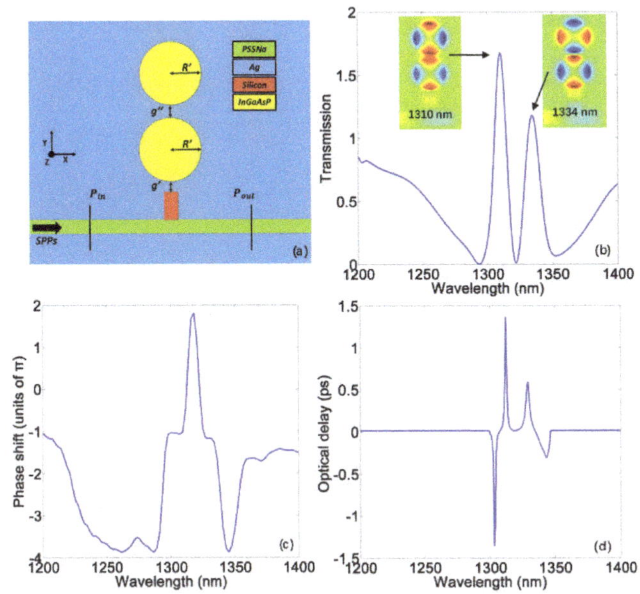

Figure 7. (**a**) 2D scheme of the gain-assisted double PIT system from the view of z-axis. The geometry parameters are g' = 28 nm, g'' = 36 nm, R' = 149 nm; (**b**) The transmission spectrum of double-disk system under the gain level $\eta = 782\ \text{cm}^{-1}$. The insets are field distributions corresponding to the two central transparent wavelengths; (**c**) The phase shift spectrum of corresponding double PIT; (**d**) The delay time of corresponding double PIT.

6. Conclusions

In summary, we report a gain-assisted plasmonic MDM system with a superior slow light performance based on PIT effect. Both the transmission and optical delay of slow light can be dramatically enhanced by the gain power. Finally, a double-channel, enhanced slow light can be achieved by introducing another disk cavity. After applying the gain material, the device performance can be enhanced and we can also realize the active control of the PIT peak and delay time. To propose such a structure, Ag film can be deposited on the Ni surface by thermal evaporation method [31], and the bus waveguide and cavities can be etched on the Ag layer by focused ion beam (FIB) [9]. Such a system also has potential applications for bio-sensing, nanolaser and optical switching.

Author Contributions: Conceptualization by Z.Z.; data curation by Z.Z.; formal analysis by Z.Z.; investigation by Z.Z.; resources by J.Z.; software by J.H. and D.C.; validation by J.Y., X.H., and Y.H.; visualization by S.X.; writing of original draft by Z.Z.; writing review and editing by J.Y.

Funding: This research was funded by the National Natural Science Foundation of China (60907003, 61671455), the Foundation of NUDT (JC13-02-13, ZK17-03-01), the Hunan Provincial Natural Science Foundation of China (13JJ3001), and the Program for New Century Excellent Talents in University (NCET-12-0142).

Conflicts of Interest: The authors declare no conflict of interest.

References

1. Fleischhauer, M.; Physik, F.; Kaiserslautern, D.; Imamoglu, A.; Marangos, J.P. Electromagnetically induced transparency: Optics in coherent media. *Rev. Mod. Phys.* **2005**, *77*, 633–673. [CrossRef]
2. Hau, L.V.; Harris, S.E.; Dutton, Z.; Behroozi, C.H. Light speed reduction to 17 meters per second in an ultracold atomic gas. *Nature* **1999**, *397*, 594–598. [CrossRef]
3. Schmidt, H.; Imamogdlu, A. Giant Kerr nonlinearities obtained by electromagnetically induced transparency. *Opt. Lett.* **1996**, *21*, 1936. [CrossRef] [PubMed]
4. Zhang, S.; Genov, D.A.; Wang, Y.; Liu, M.; Zhang, X. Plasmon-induced transparency in metamaterials. *Phys. Rev. Lett.* **2008**, *101*, 047401. [CrossRef] [PubMed]
5. Lu, H.; Liu, X.; Mao, D.; Gong, Y.; Wang, G. Induced transparency in nanoscale plasmonic resonator systems. *Opt. Lett.* **2011**, *36*, 3233–3235. [CrossRef] [PubMed]
6. Zhang, Z.; Yang, J.; He, X.; Zhang, J.; Huang, J.; Chen, D.; Han, Y. Plasmonic Refractive Index Sensor with High Figure of Merit Based on Concentric-Rings Resonator. *Sensors* **2018**, *18*, 116. [CrossRef] [PubMed]
7. Zhan, S.; Li, H.; He, Z.; Li, B.; Chen, Z.; Xu, H. Sensing analysis based on plasmon induced transparency in nanocavity-coupled waveguide. *Opt. Express* **2015**, *23*, 20313–20320. [CrossRef] [PubMed]
8. Khunsin, W.; Dorfmüller, J.; Esslinger, M.; Vogelgesang, R.; Rockstuhl, C.; Etrich, C.; Kern, K. Quantitative and direct near-field analysis of plasmonic-induced transparency and the observation of a plasmonic breathing mode. *ACS Nano* **2016**, *10*, 2214–2224. [CrossRef] [PubMed]
9. Yang, X.; Hu, X.; Chai, Z.; Lu, C.; Yang, H.; Gong, Q. Tunable ultracompact chip-integrated multichannel filter based on plasmon-induced transparencies. *Appl. Phys. Lett.* **2014**, *104*, 221114. [CrossRef]
10. Zhan, S.; Li, H.; Cao, G.; He, Z.; Li, B.; Xu, H. Theoretical analysis of plasmon-induced transparency in ring-resonators coupled channel drop filter systems. *Plasmonics* **2014**, *9*, 1431–1437. [CrossRef]
11. Wang, J.; Yuan, B.; Fan, C.; He, J.; Ding, P.; Xue, Q.; Liang, E. A novel planar metamaterial design for electromagnetically induced transparency and slow light. *Opt. Express* **2013**, *21*, 25159–25166. [CrossRef] [PubMed]
12. Lu, H.; Liu, X.; Mao, D. Plasmonic analog of electromagnetically induced transparency in multi-nanoresonator-coupled waveguide systems. *Phys. Rev. A* **2012**, *85*, 53803. [CrossRef]
13. Han, X.; Wang, T.; Li, X.; Liu, B.; He, Y.; Tang, J. Dynamically tunable slow light based on plasmon induced transparency in disk resonators coupled MDM waveguide system. *J. Phys. D Appl. Phys.* **2015**, *48*, 235102. [CrossRef]
14. Lai, G.; Liang, R.; Zhang, Y.; Bian, Z.; Yi, L.; Zhan, G.; Zhao, R. Double plasmonic nanodisks design for electromagnetically induced transparency and slow light. *Opt. Express* **2015**, *23*, 6554–6561. [CrossRef] [PubMed]
15. Akhavan, A.; Ghafoorifard, H.; Abdolhosseini, S.; Habibiyan, H. Plasmon-induced transparency based on a triangle cavity coupled with an ellipse-ring resonator. *Appl. Opt.* **2017**, *56*, 9556–9563. [CrossRef] [PubMed]
16. Lu, Q.; Wang, Z.; Huang, Q.; Jiang, W.; Wu, Z.; Wang, Y.; Xia, J. Plasmon-induced transparency and high-performance slow light in a plasmonic single-mode and two-mode resonators coupled system. *J. Lightwave Technol.* **2017**, *35*, 1710–1717. [CrossRef]
17. Kurter, C.; Tassin, P.; Zhang, L.; Koschny, T.; Zhuravel, A.P.; Ustinov, A.V.; Anlage, S.M.; Soukoulis, C.M. Classical analogue of electromagnetically induced transparency with a metal-superconductor hybrid metamaterial. *Phys. Rev. Lett.* **2011**, *107*, 043901. [CrossRef] [PubMed]
18. He, Z.; Li, H.; Zhan, S.; Li, B.; Chen, Z.; Xu, H. Tunable multi-switching in plasmonic waveguide with Kerr nonlinear resonator. *Sci. Rep.* **2015**, *5*, 15837. [CrossRef] [PubMed]
19. Liu, T.; Yi, Z.; Xiao, S. Active control of near-field coupling in a terahertz metal-graphene metamaterial. *IEEE Photonics Technol. Lett.* **2017**, *29*, 1998–2001. [CrossRef]
20. Xiao, S.; Wang, T.; Liu, T.; Yan, X.; Li, Z.; Xu, C. Active modulation of electromagnetically induced transparency analogue in terahertz hybrid metal-graphene metamaterials. *Carbon* **2017**, *126*, 271–278. [CrossRef]
21. Xiao, S.; Wang, T.; Jiang, X.; Yan, X.; Cheng, L.; Wang, B.; Xu, C. Strong interaction between graphene layer and fano resonance in terahertz metamaterials. *J. Phys. D Appl. Phys.* **2017**, *50*, 195101. [CrossRef]

22. Zheludev, N.I.; Prosvirnin, S.L.; Papasimakis, N.; Fedotov, V.A. Lasing spaser. *Nat. Photonics* **2008**, *2*, 351–354. [CrossRef]
23. Deng, Z.L.; Dong, J.W. Lasing in plasmon-induced transparency nanocavity. *Opt. Express* **2013**, *21*, 20291–20302. [CrossRef] [PubMed]
24. Luo, L.; Ge, C.; Tao, Y.; Zhu, L.; Zheng, K.; Wang, W.; Sun, Y.; Shen, P.; Guo, Z. High-efficiency refractive index sensor based on the metallic nanoslit arrays with gain-assisted materials. *Nanophotonics* **2016**, *5*, 548–555. [CrossRef]
25. Dong, Z.G.; Liu, H.; Cao, J.X.; Li, T.; Wang, S.M.; Zhu, S.N.; Zhang, X. Enhanced sensing performance by the plasmonic analog of electromagnetically induced transparency in active metamaterials. *Appl. Phys. Lett.* **2010**, *97*, 114101. [CrossRef]
26. He, J.; Wang, J.; Ding, P.; Fan, C.; Liang, E. Gain-assisted plasmon induced transparency in t-shaped metamaterials for slow light. *J. Opt.* **2015**, *17*, 055002. [CrossRef]
27. Liu, B.; Zhu, W.; Gunapala, S.D.; Stockman, M.I.; Premaratne, M. Open resonator electric spaser. *ACS Nano* **2017**, *11*, 12573–12582. [CrossRef] [PubMed]
28. Gramotnev, D.K.; Bozhevolnyi, S.I. Plasmonics beyond the diffraction limit. *Nat. Photonics* **2010**, *4*, 83–91. [CrossRef]
29. Baffou, G.; Quidant, R. Thermo-plasmonics: Using metallic nanostructures as nano-sources of heat. *Laser Photonics Rev.* **2013**, *7*, 171–187. [CrossRef]
30. Vyshnevyy, A.A.; Fedyanin, D.Y. Self-heating and cooling of active plasmonic waveguides. *ACS Photonics* **2016**, *3*, 51–57. [CrossRef]
31. Kamada, S.; Okamoto, T.; El-Zohary, S.E.; Haraguchi, M. Design optimization and fabrication of Mach-Zehnder interferometer based on MIM plasmonic waveguides. *Opt. Express* **2016**, *24*, 16224–16231. [CrossRef] [PubMed]
32. Veronis, G.; Fan, S. Modes of subwavelength plasmonic slot waveguides. *J. Lightwave Technol.* **2007**, *25*, 2511–2521. [CrossRef]
33. Yu, Z.; Veronis, G.; Fan, S.; Brongersma, M.L. Gain-induced switching in metal-dielectric-metal plasmonic waveguides. *Appl. Phys. Lett.* **2008**, *92*, 041117. [CrossRef]
34. Bamiedakis, N.; Beals, J.; Penty, R.V.; White, I.H.; Degroot, J.V.; Clapp, T.V. Cost-effective multimode polymer waveguides for high-speed on-board optical interconnects. *IEEE J. Quantum Electron.* **2009**, *45*, 415–424. [CrossRef]
35. Beadie, G.; Brindza, M.; Flynn, R.A.; Rosenberg, A.; Shirk, J.S. Refractive index measurements of poly (methyl methacrylate) (PMMA) from 0.4–1.6 μm. *Appl. Opt.* **2015**, *54*, F139–F143. [CrossRef] [PubMed]
36. Maier, S.A. *Plasmonics: Fundamentals and Applications*; Springer: Berlin, Germany, 2014; Volume 52, pp. 49–74.
37. Luo, S.; Li, B.; Xiong, D.; Zuo, D.; Wang, X. A high performance plasmonic sensor based on metal-insulator-metal waveguide coupled with a double-cavity structure. *Plasmonics* **2016**, *12*, 1–5. [CrossRef]
38. Ciminelli, C.; Campanella, C.E.; Dell'Olio, F.; Armenise, M.N. Fast light generation through velocity manipulation in two vertically-stacked ring resonators. *Opt. Express* **2010**, *18*, 2973–2986. [CrossRef] [PubMed]
39. Bigelow, M.S.; Lepeshkin, N.N.; Boyd, R.W. Superluminal and slow light propagation in a room-temperature solid. *Science* **2003**, *301*, 200–202. [CrossRef] [PubMed]
40. Ambati, M.; Nam, S.H.; Ulinavila, E.; Genov, D.A.; Bartal, G.; Zhang, X. Observation of stimulated emission of surface plasmon polaritons. *Nano Lett.* **2008**, *8*, 3998–4001. [CrossRef] [PubMed]
41. De, L.A.; Grzelczak, M.P.; Pastorizasantos, I.; Lizmarzán, L.M.; La Deda, M.; Striccoli, M.; Strangi, G. Dispersed and encapsulated gain medium in plasmonic nanoparticles: A multipronged approach to mitigate optical losses. *ACS Nano* **2011**, *5*, 5823–5829.
42. Fujita, M.; Teshima, K.; Baba, T. Low-threshold continuous-wave lasing in photopumped GaInAsP microdisk lasers: Optics and quantum electronics. *Jpn. J. Appl. Phys.* **2001**, *40*, L875–L877. [CrossRef]
43. Watanabe, T.; Saijo, Y.; Hasegawa, Y.; Watanabe, K.; Nishijima, Y.; Baba, T. Ion-sensitive photonic-crystal nanolaser sensors. *Opt. Express* **2017**, *25*, 24469–24479. [CrossRef] [PubMed]
44. Shahamat, Y.; Vahedi, M. Plasmon-induced transparency in a rectangle cavity and an H-shaped structure for sensing and switching applications. *J. Nanophotonics* **2017**, *11*, 046012. [CrossRef]

45. Kirstaedter, N.; Schmidt, O.G.; Ledentsov, N.N.; Bimberg, D.; Ustinov, V.M.; Egorov, A.Y.; Zhukov, A.E.; Maximov, M.V.; Kop'ev, P.S.; Alferov, Z.I. Gain and differential gain of single layer InAs/GaAs quantum dot injection lasers. *Appl. Phys. Lett.* **1996**, *69*, 1226–1228. [CrossRef]
46. Carrere, H.; Marie, X.; Lombez, L.; Amand, T. Optical gain of INGAASN/INP quantum wells for laser applications. *Appl. Phys. Lett.* **2006**, *89*, 181115. [CrossRef]

materials

MDPI

Article

Independently Tunable Fano Resonances Based on the Coupled Hetero-Cavities in a Plasmonic MIM System

Qiong Wang [1,2], Zhengbiao Ouyang [1,2,*], Mi Lin [1,2] and Qiang Liu [1,2]

1 THz Technical Research Center of Shenzhen University, Shenzhen University, Shenzhen 518060, China; qwang@szu.edu.cn (Q.W.); linfengas111@szu.edu.cn (M.L.); qliu@szu.edu.cn (Q.L.)

2 College of Electronic Science &Technology, Shenzhen University, Shenzhen 518060, China

* Correspondence: zbouyang@szu.edu.cn; Tel.: +86-755-2653-4230

Received: 17 August 2018; Accepted: 6 September 2018; Published: 10 September 2018

Abstract: In this paper, based on coupled hetero-cavities, multiple Fano resonances are produced and tuned in a plasmonic metal-insulator-metal (MIM) system. The structure comprises a rectangular cavity, a side-coupled waveguide, and an upper-coupled circular cavity with a metal-strip core, used to modulate Fano resonances. Three Fano resonances can be realized, which originate from interference of the cavity modes between the rectangular cavity and the metal-strip-core circular cavity. Due to the different cavity-cavity coupling mechanisms, the three Fano resonances can be divided into two groups, and each group of Fano resonances can be well tuned independently by changing the different cavity parameters, which can allow great flexibility to control multiple Fano resonances in practice. Furthermore, through carefully adjusting the direction angle of the metal-strip core in the circular cavity, the position and lineshape of the Fano resonances can be easily tuned. Notably, reversal asymmetry takes place for one of the Fano resonances. The influence of the direction angle on the figure of merit (FOM) value is also investigated. A maximum FOM of 3436 is obtained. The proposed structure has high transmission, sharp Fano lineshape, and high sensitivity to change in the background refractive index. This research provides effective guidance to tune multiple Fano resonances, which has important applications in nanosensors, filters, modulators, and other related plasmonic devices.

Keywords: tunable fano resonances; surface plasmon polaritons; coupled cavities; finite element method

1. Introduction

Noble metallic nanostructures that support surface plasmon polaritons (SPPs) have stimulated tremendous research interest due to their special capabilities of overcoming traditional optical diffraction limits, controlling light in the nanoscale domain, and producing extremely strong local electromagnetic fields [1–5]. They provide the possibility for devices with extraordinary properties, high-degree miniaturization, and large-scale integration. As we know, due to electromagnetic wave interactions with the metal surfaces in SPP devices, the transmission characteristics in the devices are closely related to the shape and size of the designed geometry. By now, various plasmonic devices have been proposed, such as optical filters, wavelength division multiplexers, and information modulators [6,7], which open up new opportunities for fabricating plasmonic integrated on-chip systems.

On the other hand, Fano resonance [8–14] has emerged as a new research aspect in SPP devices, since it exhibits sharp and asymmetric lineshape in spectra, and small perturbations can induce dramatic intensity variation and wavelength shift. Fano resonance arises from the coherent coupling and interference between a discrete state (or a narrow spectrum) and a continuous state (or a broad spectrum), so that unique spectral patterns can be produced [15–17]. Recently, considerable effort has been devoted to research on the tunability of Fano resonance, as it has important applications in nanosensors, filters, slow-light devices, modulators, and so on [18–20]. For example, it has been reported that a type of E-shaped plasmonic nanostructure was formed by the interference of the quadrupole resonance modes from a C-shaped metal ring with the dipolar resonance modes from a metal strip, in which the tunability can be easily realized by changing the asymmetry of the geometry [21]. In addition, a compact plasmonic sensor has been designed with a stub and a side-coupled split-ring resonator, and the wavelength position and Fano lineshape can be adjusted by changing the opening direction of the split ring [22]. Additionally, a kind of electrically tunable Fano-type resonance of asymmetric metal-wire pairs has been achieved by controlling the varactor diode loaded on the plasmonic device [23].

For some specific plasmonic structures, multiple Fano resonances can be obtained [24,25], but the independent tunability has rarely been reported. The research on this topic should be given more attention since it will bring great flexibility for fabricating multiple-function devices.

In this paper, based on coupled hetero-cavities, independently tunable Fano resonances are produced and investigated in a plasmonic metal-insulator-metal (MIM) system. This consists of a rectangular cavity, a side-coupled waveguide, and an upper-coupled circular cavity with an angle-tunable metal strip in the center. Through investigating the coupling effect of the two above-mentioned cavities, the mechanisms of Fano resonances are explored. Due to the different cavity-cavity coupling effects, independent tunability can be realized by changing different cavity parameters. Furthermore, the influence of the direction angle of the metal-strip core on Fano resonances is investigated. The figure of merit (FOM) values for different direction angles are also calculated and compared. This research provides an affective measure to produce multiple Fano resonances and realize independent tunability.

2. Structure Design

The X-Y top view of the designed plasmonic nanosystem based on coupled hetero-cavities is shown in Figure 1a. Considering the calculation time, a two-dimensional model is used to demonstrate the characteristics of the structure. It is necessary to emphasize that the MIM structure is chosen because it has the remarkable advantages of long propagation distance, deep-subwavelength field confinement, low bend loss, and easy integration [26–29].

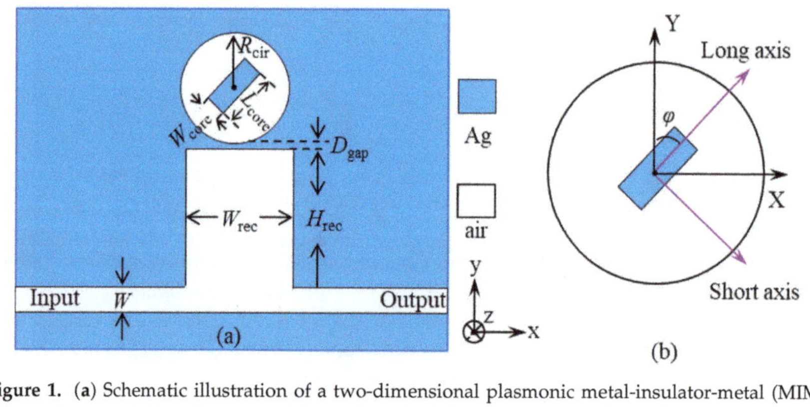

Figure 1. (**a**) Schematic illustration of a two-dimensional plasmonic metal-insulator-metal (MIM) nanosystem consisting of a rectangular cavity coupled with a waveguide, and an upper-coupled circular cavity with an angle-tunable metal strip located at the center. (**b**) In the metal-strip-core circular cavity, the rotation direction of the metal strip is defined by an angle φ between the y-axis and the long axis of the metal strip.

As shown in Figure 1a, a rectangular cavity is connected to a MIM waveguide. The width and height of the rectangular cavity are denoted as W_{rec} and H_{rec}, respectively. The waveguide has a width of W = 65 nm. Above the rectangular cavity it has an upper-coupled circular air cavity with a metal-strip core in the center. The radius of the circular cavity is denoted as R_{cir}. The vertical gap between the circular and rectangular cavities is set as D_{gap} = 10 nm. L_{core} and W_{core} represent the length and width of the metal strip, respectively. In order to modulate Fano resonances, the rotation direction of the metal strip is adjusted. An angle φ between the y-axis and the long axis of the metal strip is defined, as shown in Figure 1b. The blue and white areas denote the noble metal of silver and air, respectively. For silver, its frequency-dependent complex relative permittivity is characterized by the Drude model [30,31]:

$$\varepsilon_m(\omega) = \varepsilon_\infty - \frac{\omega_p^2}{\omega(\omega + i\gamma)} \tag{1}$$

where ε_∞ is the dielectric constant at infinite frequency, γ is the electron collision frequency, ω is the frequency of the incident light, and ω_p is the bulk plasma frequency. The parameters are ε_∞ = 3.7, ω_p = 1.38×10^{16} Hz, and γ = 2.73×10^{13} Hz.

In the following section, the transmission characteristics of the coupled hetero-cavity system are numerically simulated in detail using COMSOL software (Version 5.3, Stockholm, Sweden). The structure is divided into a grid of about 5×10^4 small cells. Perfectly matched layers are added around the calculated domain to absorb the electromagnetic waves going out of the structure. A transverse magnetic (TM) wave is launched at the left waveguide. The incident power P_{in} and transmitted power P_{out} are detected by two power monitors set at the input and output ports, respectively. The total transmission of the coupled system is calculated as P_{out}/P_{in}.

3. Results and Discussion

3.1. The Coupling Mechanism of the Three Fano Resonances

As we know, Fano resonance always exhibits a sharp and asymmetric spectral pattern, which results from the coupling effect between a wide continuous state and a narrow distinct state. In this section, we will investigate the Fano resonance of the coupled hetero-cavity system. In order to explore the mechanisms of Fano resonances, it is necessary to study the rectangular cavity and the metal-strip-core circular cavity separately, and then consider their coupling effect.

Figure 2a shows the transmission of a metal-strip-core circular cavity coupled with a waveguide. The structure is plotted in the inset. The radius of the circular cavity is chosen as R_{cir} = 150 nm, and the cavity-waveguide gap is set as D_{gap} = 10 nm. For the metal strip of the circular cavity, its width, length, and direction angle are set as W_{core} = 60 nm, L_{core} = 150 nm, and φ = 30°, respectively. It can be seen that the circular cavity keeps a certain distance from the waveguide, thus strong cavity modes can only be excited at particular wavelengths. In Figure 2a, it can be seen that two very narrow transmission dips appear at the wavelengths of λ = 960 nm and λ = 737 nm, respectively. For convenience, the two modes are named as TD1 and TD2, respectively. Notably, TD1 exhibits an asymmetrical resonance shape. This can be considered as a type of simple Fano resonance that originates from the mode coupling of the metal-strip-core circular cavity and the waveguide. Meanwhile, through adjusting the direction angle φ of the metal strip core in the cavity, the coupling effect can be tuned. TD1 and TD2 correspond to two different dipole modes that are anti-symmetric about the long and short axes of the metal strip, respectively, as can be seen from the simulated H_z magnetic field distributions shown in Figure 2b,c.

Figure 2d shows the transmission of a rectangular cavity connecting with a waveguide. The width and height of the rectangular cavity are chosen as W_{rec} = 300 nm and H_{rec} = 500 nm, respectively. It is necessary to point out that the rectangular cavity is chosen to be large enough, and designed to directly connect with the waveguide, so that the cavity modes can be efficiently and strongly excited at a broad range of wavelengths. As shown in Figure 2d, we can see that two transmission peaks appear at the wavelengths of λ = 1118 nm and λ = 707 nm, which are named as TP1 and TP2, respectively. It is obvious that mode TP1 has a very wide peak, with a half-high width which is from λ = 949 nm to λ = 1323 nm. Mode TP2 has a half-high width of 25 nm, which is not as wide as mode TP1. The corresponding H_z magnetic field distributions are given in Figure 2e,f, respectively. Strong cavity modes can be observed, which are anti-symmetric about the short and long axes of the rectangular cavity, respectively.

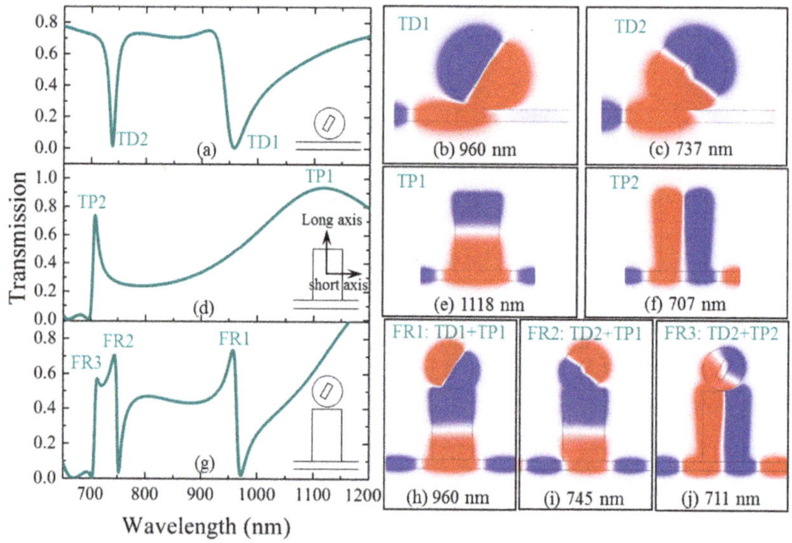

Figure 2. (**a**) The transmission of a metal-strip-core cavity coupled with a waveguide. The H_z magnetic field distributions of the transmission dips at (**b**) TD1, λ = 960 nm; (**c**) TD2, λ = 737 nm. (**d**) The transmission of a rectangular cavity connected with a waveguide. The H_z magnetic field distributions of the transmission peaks at (**e**) TP1, λ = 1118 nm; (**f**) TP2, λ = 707 nm. (**g**) The transmission of a coupled system consisting of a waveguide, a rectangular cavity, and a metal-strip-core circular cavity. The H_z magnetic field distributions of the Fano peaks at (**h**) FR1, λ = 960 nm; (**i**) FR2, λ = 745 nm; (**j**) FR3, λ = 711 nm.

Figure 2g shows the transmission of the coupled hetero-cavity system. The parameters of the two cavities are the same as those in Figure 2a,d. The cavity-cavity gap is set as D_{gap} = 10 nm. We can observe that three resonances with sharp and asymmetric spectral patterns appear at λ = 960 nm, λ = 745 nm, and λ = 711 nm, denoted by FR1, FR2, and FR3, respectively. The lineshapes have the typical characteristic of Fano resonance. The three Fano resonances in Figure 2g can be regarded as the coupling of the cavity modes from the rectangular cavity and the metal-strip-core cavity. In order to further investigate the coupling effects of the Fano resonances, the H_z magnetic field distributions for the three Fano peaks of FR1, FR2, and FR3, are simulated and illustrated in Figure 2h–j. This shows that they originate from three different types of cavity-cavity coupling mechanisms, that is, FR1, FR2, and FR3 are formed by the coupling effects of modes TD1 and TP1, modes TD2 and TP1, and modes TD2 and TP2, respectively, where TD1 and TD2 are the transmission dips of the metal-strip-core circular cavity, and TP1 and TP2 are the transmission peaks of the rectangular cavity.

3.2. The Characteristics of Independent Tunability of the Three Fano Resonances

In order to further understand the Fano resonances, the influence of the cavity parameters on transmission is investigated in detail. Figure 3a shows the change of transmission with the radius of the circular cavity increasing from R_{cir} = 155 nm to R_{cir} = 175 nm. The other parameters remain unchanged.

We can see that the Fano peaks of FR1 and FR2 have obvious red shifts, with wavelength increments of $\Delta\lambda$ = 83 nm and $\Delta\lambda$ = 74 nm, respectively, while the Fano peak of FR3 is nearly unchanged. This proves that the Fano resonances FR1 and FR2 can be simultaneously adjusted by changing the parameter of R_{cir}, and the Fano resonance FR3 is not affected.

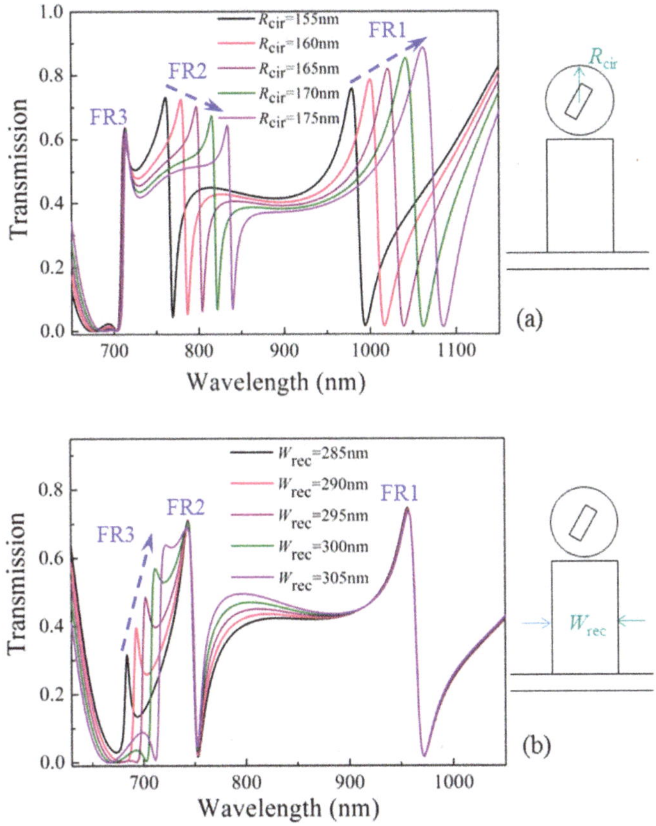

Figure 3. Transmission changes for (**a**) different radii of the circular cavity, R_{cir} = 155 nm, 160 nm, 165 nm, 170 nm, 175 nm; (**b**) different widths of the rectangular cavity, W_{rec} = 285 nm, 290 nm, 295 nm, 300 nm, 305 nm.

Figure 3b shows the change of transmission with the width of the rectangular cavity increasing from W_{rec} = 285 nm to W_{rec} = 305 nm. The other parameters remain unchanged. The result is the opposite compared with the above case. The Fano peaks of FR1 and FR2 have almost no change, while the Fano peak of FR3 exhibits a red shift of $\Delta\lambda$ = 40 nm in wavelength. This proves that the Fano resonance FR3 can be freely adjusted by changing the parameter of W_{rec}, and the other two Fano resonances, FR1 and FR2, are not affected.

From the above results, we can find an interesting phenomenon. The three Fano resonances can be divided into two groups. Fano resonances FR1 and FR2 belong to the same group (Group 1), and Fano resonance FR3 belongs to the other group (Group 2). The two groups of Fano resonances can be independently controlled by different cavity parameters. This phenomenon can be understood from the resonances of the rectangular cavity. As shown in Figure 2e,f, the input light is set to be a transverse magnetic (TM) plane wave, thus the resonance mode of the rectangular cavity can be denoted as TM_{mn}, where m and n are integers representing the resonant orders in the transverse (x-axis) and longitudinal (y-axis) directions of the rectangular cavity. The transmission peaks TP1 and TP2 can be regarded as TM_{01} (see Figure 2e) and TM_{10} (see Figure 2f) modes, respectively. For the coupled hetero-cavity system, the H_z field distribution of the rectangular cavity for FR1 or FR2 is TM_{01} (see Figure 2h,i), which is sensitive to the longitudinal parameters, such as the height of the rectangular cavity (H_{rec}), or the radius of the circular cavity (R_{cir}). Due to the strong coupling effect existing between the hetero-cavities, the longitudinal resonance of the rectangular cavity is more sensitive to the change of R_{cir} than that of H_{rec}. Therefore, R_{cir} is chosen as an example to independently control the Fano resonance FR1 (or FR2). On the other hand, the H_z field distribution of the rectangular cavity for FR3 is TM_{10} (see Figure 2j), which is sensitive to the transverse parameters, such as the width of the rectangular cavity (W_{rec}).

3.3. *Tuning Fano Resonances by Changing the Direction Angle of the Metal Strip*

In order to investigate the tunability of the nanostructure, the influence of the direction angle φ on Fano resonances is also considered. When φ is changed, the field distribution of the circular cavity varies accordingly, which can directly affect the coupling of the hetero-cavities. Thus, the Fano resonances are easily tuned. Figure 4a shows the transmissions with φ changing from 0° to 90°. We can see that when φ increases, FR1 becomes stronger and FR2 becomes weaker, both accompanied by obvious wavelength shifts, while FR3 has almost no change. The reason for the above difference between FR1, FR2, and FR3, is because φ has greater impact on the longitudinal resonance mode TM_{01} of the rectangular cavity (existing in FR1 and FR2) than that of the transverse resonance mode TM_{10} (existing in FR3).

Remarkably, we notice that when φ increases to an angle of 90°, the Fano resonance FR2 turns into the reversal Fano asymmetry, as can be seen by comparing the two cases of $\varphi = 90°$ and $\varphi = 0°$, marked by the two red squares in Figure 4a. Further verification can be obtained by calculating the q values from the Fano formula. The transmission can be expressed by the following Fano formula [32],

$$T = T_{\text{Bethe}} + C\frac{(\lambda - \lambda_{\text{Res}} + q\Gamma/2)^2}{(\lambda - \lambda_{\text{Res}})^2 + (\Gamma/2)^2} \tag{2}$$

where T_{Bethe} is the direct transmission referred to as Bethe's contribution, C is the non-resonant transmission coefficient, λ_{Res} is the resonant wavelength, Γ is the linewidth, and q is a dimensionless parameter that describes the asymmetry profile. Using the Origin software, the transmissions of Fano resonances FR1, FR2, and FR3 are fitted with Equation (2). When the fitting error reaches the minimum value, the fitted line expressed by Equation (2) can be determined. Then the variables T_{Bethe}, C, λ_{res}, q, and Γ are fixed. We investigate the dependence of the asymmetry parameter q on the direction angle φ, as shown in Figure 4b. We can find that for Fano resonance FR2, the q value varies from negative ($q = -1.15$) for $\varphi = 0°$ to positive ($q = 0.27$) for $\varphi = 90°$. This proves that the reversed asymmetry of the Fano resonance is formed. This phenomenon is attributed to the difference in the phase shifts of the cavity modes caused by adjusting the direction angle φ. The phase shift is defined as the phase difference between the input and output positions of the rectangular cavity. In Figure 4c,d, the H_z magnetic field distributions for the two cases of $\varphi = 0°$ and $\varphi = 90°$ are shown. For $\varphi = 0°$, the phase shift of the TM_{01} mode in the rectangular cavity is about 0°, while for $\varphi = 90°$, FR2 has a blue shift under the influence of field modulation induced by the rotation of the metal strip (from $\varphi = 0°$ to $\varphi = 90°$), leading to the mode in the rectangular cavity changing from TM_{01} to TM_{10}. The corresponding phase shift becomes π. As a consequence, the reversed asymmetry of the Fano resonance is generated. Furthermore, we can see that $q = 0$ is realized at $\varphi = 82°$, meaning that the resonance exhibits a symmetric profile for this case. This point can be considered as a critical condition to obtain an anti-resonant pattern, and it demonstrates an important tunability option for Fano resonance. The drastic change of the asymmetry parameter can be explored for applications such as optical switches, information modulators, and nonlinear and slow-light devices.

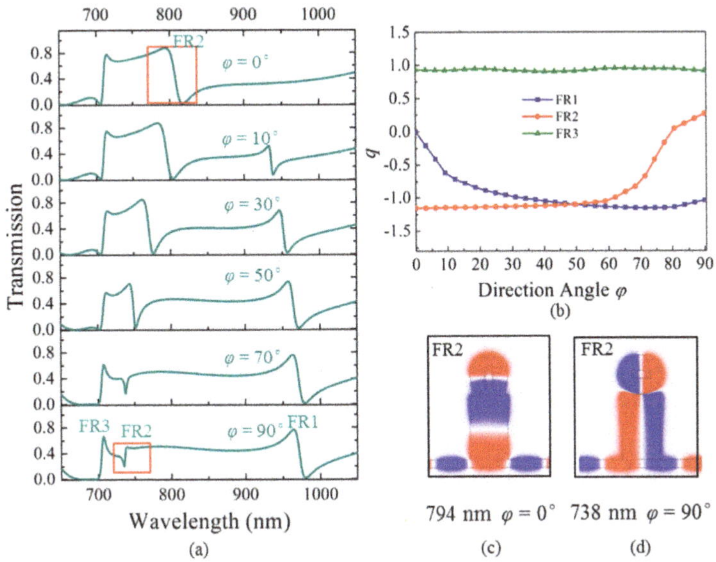

Figure 4. (**a**) Transmissions from changing the direction angle φ of the metal strip in the circular cavity with $\varphi = 0°$, $10°$, $30°$, $50°$, $70°$, and $90°$. (**b**) The q values are calculated for φ changing from $0°$ to $90°$. The H_z magnetic field distributions for the Fano peak of FR2 at (**c**) $\varphi = 0°$ and (**d**) $\varphi = 90°$.

As we know, sensing performance is one of the most important applications for Fano resonance. Here, the sensitivity characteristic of Fano resonance to the refractive index of the background dielectric material is investigated by calculating the figure of merit (FOM) value [33]. FOM is defined as the maximum of $\Delta T/(T\Delta n)$ when Δn and T are changed. $\Delta T/(T\Delta n)$ describes the relative transmission variation $\Delta T/T$ at a fixed wavelength induced by the refractive index change Δn of the dielectric material, where T denotes the transmission in the proposed structure. As shown in Figure 5, the FOM values of Fano resonances FR1, FR2, and FR3, are calculated when the direction angle φ is scanned from $0°$ to $90°$. The other parameters remain unchanged. We can see that the FOM value of FR1 becomes larger with the increase of φ at first, and then it reaches the maximum value of 3436 at $\varphi = 75°$ (λ = 978 nm at the Fano dip), after that it begins to decrease slightly. The high FOM value contributes to the sharp Fano lineshape in transmission. For FR2, the FOM value decreases with the increase of φ. This is because the intensity of the Fano resonance becomes weaker. In contrast, the FOM of FR3 has little change, with the FOM value almost keeping above 1420. Comparing Figures 4b and 5, we can see that a sharp Fano lineshape always has large $|q|$, which is beneficial to obtain a large FOM value for sensing. The proposed structure has high transmission, sharp Fano lineshape, and high sensitivity to the change in background refractive index, which has important applications in nanosensors, filters, modulators, and other related plasmonic devices.

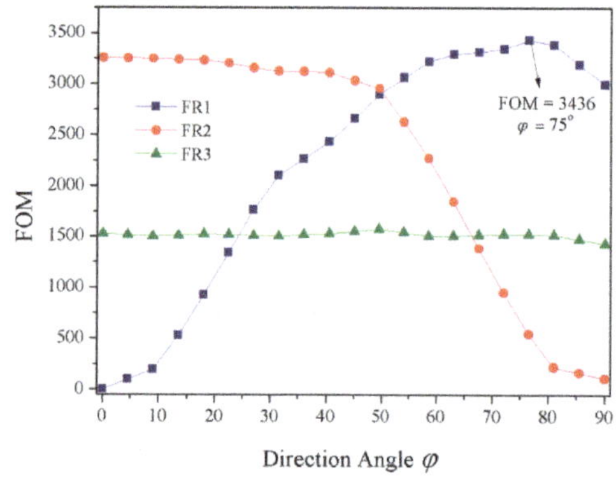

Figure 5. The figure of merit (FOM) values of the Fano resonances FR1, FR2, and FR3 related to the direction angle φ.

4. Conclusions

In summary, a type of coupled hetero-cavity structure is proposed to produce tunable Fano resonances in a plasmonic MIM system. It consists of a rectangular cavity, a side-coupled waveguide, and an upper-coupled circular cavity with a metal-strip core used to modulate Fano resonances. Three Fano resonances can be realized, which originate from the interference of the modes of the rectangular cavity and the metal-strip-core circular cavity. Due to the different cavity-cavity coupling mechanisms, the three Fano resonances can be divided into two groups, and each group of Fano resonances can be well independently tuned by changing the different cavity parameters, which has the advantage of flexibly modulating Fano resonances. Furthermore, through carefully adjusting the direction angle of the metal-strip core in the circular cavity, the position and lineshape of the Fano resonances can be easily tuned. Notably, reversal asymmetry takes place for one of the Fano resonances when the direction angle of the metal strip is changed to be $\varphi = 90°$. The results show that a maximum FOM value of 3436 is obtained. This research provides effective guidance to produce and tune multiple Fano resonances, which has important applications in nanosensors, filters, slow-light devices, modulators, and other related plasmonic devices.

Author Contributions: Q.W. designed the structure and wrote the paper; M.L. analyzed the theory; Q.L. performed the simulations; Z.O. analyzed the data and revised the whole paper.

Funding: This work was financially supported by the National Natural Science Foundation of China (NSFC) (No. 61605128, 61275043, 61307048, 11404220), the Natural Science Foundation of Guangdong Province (No. 2017A030310455), and Start-up Funds for New Teachers of Shenzhen University (No. 2016024).

Conflicts of Interest: The authors declare no conflicts of interest.

References

1. Caldwell, J.D.; Lindsay, L.; Giannini, V.; Vurgaftman, I.; Reinecke, T.L.; Maier, S.A.; Glembocki, O.J. Low-loss, infrared and terahertz nanophotonics using surface phonon polaritons. *Nanophotonics* **2015**, *4*, 44–68. [CrossRef]
2. Lin, J.; Balthasar Mueller, J.P.; Wang, Q.; Yuan, G.H.; Antoniou, N.; Yuan, X.C.; Capasso, F. Polarization-controlled tunable directional coupling of surface plasmon polaritons. *Science* **2013**, *340*, 331–334. [CrossRef] [PubMed]
3. Williams, C.R.; Andrews, S.R.; Maier, S.A.; Fernandez-Dominguez, A.I.; Martin-Moreno, L.; Garcia-Vidal, F.J. Highly confined guiding of terahertz surface plasmon polaritons on structured metal surfaces. *Nat. Photonics* **2008**, *2*, 175–179. [CrossRef]
4. Brawley, Z.T.; Bauman, S.J.; Darweesh, A.A.; Debu, D.T.; Ladani, F.T.; Herzog, J.B. Plasmonic Au array SERS substrate with optimized thin film oxide substrate layer. *Materials* **2018**, *11*, 942. [CrossRef] [PubMed]

5. Rifat, A.A.; Rahmani, M.; Xu, L.; Miroshnichenko, A.E. Hybrid metasurface based tunable near-perfect absorber and plasmonic sensor. *Materials* **2018**, *11*, 1091. [CrossRef] [PubMed]
6. Zayats, A.V.; Smolyaninov, I.I.; Maradudin, A.A. Nano-optics of surface plasmon polaritons. *Phys. Rep.* **2005**, *408*, 131–314. [CrossRef]
7. Zhou, S.Y.; Lin, J.Y.; Wong, S.W.; Deng, F.; Zhu, L.; Yang, Y.; He, Y.J.; Tu, Z.H. Spoof surface plasmon polaritons power divider with large Isolation. *Sci. Rep.* **2018**, *8*, 5947. [CrossRef] [PubMed]
8. Giannini, V.; Francescato, Y.; Amrania, H.; Phillips, C.C.; Maier, S.A. Fano resonances in nanoscale plasmonic systems: A parameter-free modeling approach. *Nano Lett.* **2011**, *11*, 2835–2840. [CrossRef] [PubMed]
9. Lassiter, J.B.; Sobhani, H.; Knight, M.W.; Mielczarek, W.S.; Nordlander, P.; Halas, N.J. Designing and deconstructing the Fano lineshape in plasmonic nanoclusters. *Nano Lett.* **2012**, *12*, 1058–1062. [CrossRef] [PubMed]
10. Lukyanchuk, B.; Zheludev, N.I.; Maier, S.A.; Halas, N.J.; Nordlander, P.; Giessen, H.; Chong, C.T. The Fano resonance in plasmonic nanostructures and metamaterials. *Nat. Mater.* **2010**, *9*, 707–715. [CrossRef] [PubMed]
11. Limonov, M.F.; Rybin, M.V.; Poddubny, A.N.; Kivshar, Y.S. Fano resonances in photonics. *Nat. Photonics* **2017**, *11*, 543–554. [CrossRef]
12. Zafar, R.; Salim, M. Analysis of asymmetry of Fano resonance in plasmonic metal-insulator-metal waveguide. *Photonics Nanostruct. Fundam. Appl.* **2017**, *23*, 1–6. [CrossRef]
13. Li, S.L.; Wang, Y.L.; Jiao, R.Z.; Wang, L.L.; Duan, G.Y.; Yu, L. Fano resonances based on multimode and degenerate mode interference in plasmonic resonator system. *Opt. Express* **2017**, *25*, 3525–3533. [CrossRef] [PubMed]
14. Chen, J.J.; Sun, C.W.; Gong, Q.H. Fano resonances in a single defect nanocavity coupled with a plasmonic waveguide. *Opt. Lett.* **2014**, *39*, 52–55. [CrossRef] [PubMed]
15. Miroshnichenko, A.E.; Kivshar, Y.S.; Vicencio, R.A.; Molina, M.I. Fano resonance in quadratic waveguide arrays. *Opt. Lett.* **2005**, *30*, 872–874. [CrossRef] [PubMed]
16. Kobayashi, K.; Aikawa, H.; Sano, A.; Katsumoto, S.; Iye, Y. Fano resonance in a quantum wire with a side-coupled quantum dot. *Phys. Rev. B* **2004**, *70*, 035319. [CrossRef]
17. Miroshnichenko, A.E.; Flach, S.; Kivshar, Y.S. Fano resonances in nanoscale structures. *Rev. Mod. Phys.* **2010**, *82*, 2257. [CrossRef]
18. Qi, J.W.; Chen, Z.Q.; Chen, J.; Li, Y.D.; Qiang, W.; Xu, J.J.; Sun, Q. Independently tunable double Fano resonances in asymmetric MIM waveguide structure. *Opt. Express* **2014**, *22*, 14688–14695. [CrossRef] [PubMed]
19. Deng, Y.; Cao, G.T.; Yang, H.; Li, G.H.; Chen, X.S.; Lu, W. Tunable and high-sensitivity sensing based on Fano resonance with coupled plasmonic cavities. *Sci. Rep.* **2017**, *7*, 10639. [CrossRef] [PubMed]
20. Zhan, S.P.; Peng, Y.Y.; He, Z.H.; Li, B.X.; Chen, Z.Q.; Xu, H.; Li, H.J. Tunable nanoplasmonic sensor based on the asymmetric degree of Fano resonance in MDM waveguide. *Sci. Rep.* **2016**, *6*, 22428. [CrossRef] [PubMed]
21. Ren, X.B.; Ren, K.; Cai, Y.X. Tunable compact nanosensor based on Fano resonance in a plasmonic waveguide system. *Appl. Opt.* **2017**, *56*, H1–H9. [CrossRef] [PubMed]
22. Sun, B.; Zhao, L.X.; Wang, C.; Yi, X.Y.; Liu, Z.Q.; Wang, G.H.; Li, J.M. Tunable Fano resonance in E-shape plasmonic nanocavities. *J. Phys. Chem. C* **2014**, *118*, 25124–25131. [CrossRef]
23. Fu, Q.H.; Zhang, F.L.; Fan, Y.C.; He, X.; Qiao, T.; Kong, B.T. Electrically tunable Fano-type resonance of an asymmetric metal wire pair. *Opt. Express* **2016**, *24*, 11708–11715. [CrossRef] [PubMed]
24. Chen, Z.; Song, X.K.; Duan, G.Y.; Wang, L.L.; Yu, L. Multiple Fano resonances control in MIM side-coupled cavities systems. *IEEE Photonics J.* **2015**, *7*, 2701009. [CrossRef]
25. Zhang, B.H.; Wang, L.L.; Li, H.J.; Zhai, X.; Xia, S.X. Two kinds of double Fano resonances induced by an asymmetric MIM waveguide structure. *J. Opt.* **2016**, *18*, 065001. [CrossRef]
26. Taheri, A.N.; Kaatuzian, H. Numerical investigation of a nano-scale electro-plasmonic switch based on metal-insulator-metal stub filter. *Opt. Quantum Electron.* **2015**, *47*, 159–168. [CrossRef]
27. Rezaei, M.; Jalaly, S.; Miri, M.; Khavasi, A.; Fard, A.P.; Mehrany, K.; Rashidian, B. A distributed circuit model for side-coupled nanoplasmonic structures with metal-insulator-metal arrangement. *IEEE J. Sel. Top. Quantum Electron.* **2012**, *18*, 1692–1699. [CrossRef]
28. Taheri, A.N.; Kaatuzian, H. Design and simulation of a nanoscale electroplasmonic 1 × 2 switch based on asymmetric metal-insulator-metal stub filters. *Appl. Opt.* **2014**, *53*, 6546–6553. [CrossRef] [PubMed]

29. Ogawa, S.; Kimata, M. Metal-insulator-metal-based plasmonic metamaterial absorbers at visible and infrared wavelengths: A review. *Materials* **2018**, *11*, 458. [CrossRef] [PubMed]
30. Chen, Z.; Hu, R.; Cui, L.N.; Yu, L.; Wang, L.L.; Xiao, J.H. Plasmonic wavelength demultiplexers based on tunable Fano resonance in coupled-resonator systems. *Opt. Commun.* **2014**, *320*, 6–11. [CrossRef]
31. Jankovic, N.; Cselyuszka, N. Multiple Fano-like MIM plasmonic structure based on triangular resonator for refractive index sensing. *Sensors* **2018**, *18*, 287. [CrossRef] [PubMed]
32. Genet, C.; Van Exter, M.P.; Woerdman, J.P. Fano-type interpretation of red shifts and red tails in hole array transmission spectra. *Opt. Commun.* **2003**, *225*, 331–336. [CrossRef]
33. Wen, K.H.; Hu, Y.H.; Chen, L.; Zhou, J.Y.; Lei, L.; Guo, Z. Fano resonance with ultra-high figure of merits based on plasmonic metal-insulator-metal waveguide. *Plasmonics* **2015**, *10*, 27–32. [CrossRef]

materials

MDPI

Article

Miniature Spectroscopes with Two-Dimensional Guided-Mode Resonant Metal Grating Filters Integrated on a Photodiode Array

Yoshiaki Kanamori *, Daisuke Ema and Kazuhiro Hane

Department of Finemechanics, Tohoku University, Miyagi 980-8579, Japan; ron@hane.mech.tohoku.ac.jp (D.E.); hane2@hane.mech.tohoku.ac.jp (K.H.)

* Correspondence: kanamori@hane.mech.tohoku.ac.jp; Tel.: +81-22-795-6965

Received: 18 September 2018; Accepted: 5 October 2018; Published: 10 October 2018

Abstract: A small spectroscope with 25 color sensors was fabricated by combining metamaterial color filters and Si photodiodes. The metamaterial color filters consisted of guided-mode resonant metal gratings with subwavelength two-dimensional periodic structures. Transmittance characteristics of the color filters were designed to obtain peak wavelengths proportional to grating periods. For each color sensor, a peak wavelength of the spectral sensitivity could be tuned in the range of visible wavelengths by adjusting each grating period. By performing spectrum reconstruction using Tikhonov regularization, the spectrum of an incident light was obtained from the signal of photodiodes. Several monochromatic lights were made incident on the fabricated device and the spectral characteristics of the incident light were reconstructed from the output signals obtained from the respective color sensors. The peak wavelengths of the reconstructed spectra were in good agreement with the center wavelengths of the monochromatic lights.

Keywords: spectroscopes; metamaterial; plasmonics; structural color filters; photodiodes

1. Introduction

Color filters, which work as wavelength selective filters, have been used in image sensors and liquid crystal displays. In recent years, since the discovery of extraordinary transmission phenomenon based on surface plasmon by Ebbesen et al. in 1998 [1], plasmonic color filters using metal nanostructures have been actively studied [2–6]. Plasmonic color filters have various advantages over the conventional color filters using pigment, such that various color characteristics can be realized depending on the structural shapes with thicknesses of just tens of nanometers. Therefore, plasmonic color filters for many colors can be fabricated on the same substrate by a single fabrication process, unlike the conventional color filters using pigment. Also, plasmonic color filters have high compatibility with complementary metal–oxide–semiconductor and charge-coupled devices based on semiconductor microfabrication technologies, compared with color filters using conventional pigments. Furthermore, in recent years, a nanoimprint technology in which submicron structures are easily formed using molds has been advanced [7–18], and improved productivity of plasmonic nanostructures can be expected.

Spectroscopes are widely used [19] as instruments for measuring the energy intensity for each wavelength of light. Since most spectroscopes use diffraction gratings as wavelength selective elements, a certain propagation distance is required to separate diffracted waves spatially. Therefore, it is difficult to miniaturize the spectroscopes that use diffraction gratings. Moreover, precise optical axis adjustment is required for assembling the numerous optical components in the construction of the spectroscopes, which results in high cost. In industrial fields, colorimeters using spectroscopes are used as devices for measuring color. To meet the increasing demand for spectroscopic devices, such as the color management of products and food quality control, cost reduction and the downsizing of spectroscopic devices are strongly required.

In recent years, novel spectroscopes fabricated by combining plasmonic color filter array and photodiode array have been reported [20–26]. A color filter is formed on each photodiode, and spectral information is calculated using output signals obtained from photodiodes. Such filter array spectroscopes do not require long propagation distances like the ones using diffraction gratings and are thought to have the possibility of expanding the range of use of spectroscopic devices as ultrasmall and inexpensive spectroscopes. However, the wavelength selectivity of color filters used for the filter array spectroscopes reported are insufficient. Instead, increasing the number of filters and implementation of an improved calculation algorithm have been attempted to improve the spectroscopic characteristics of devices [22]. Increasing the number of filters leads to an increase in the light receiving area, contrary to miniaturization. Also, complicated calculation processing leads to the load and delay of calculation processing. Therefore, to improve filter array spectroscope devices, the improvement of the filter characteristics itself is required. Although guided-mode resonant gratings are known as high-efficiency wavelength selective filters [27–36], they function as reflection type wavelength selective filters. Transmission type wavelength selective filters are necessary for application in filter array spectroscopes.

As mentioned above, various plasmonic color filters have been reported so far. However, as far as we know, there are only few reports on filter array spectroscopes integrating color filters and photodiodes. Therefore, research on the improvement of their characteristics is required. In this study, we design and fabricate a filter array spectroscope and evaluate the characteristics of the fabricated spectroscopic device integrating newly designed color filters and photodiode array. There are no reports of filter array spectroscopes combining guided-mode resonant gratings and photodiodes, as far as we know. We have newly designed transmission type wavelength selective filters based on guided-mode resonant metal gratings with subwavelength two-dimensional (2D) periodic structures without polarization dependency for normal incident light.

2. Device Configuration

Figure 1a shows a conceptual diagram of the proposed spectroscopic device. A plurality of color sensors with different spectral sensitivity characteristics, in which metamaterial color filters having different spectral characteristics are stacked and arranged on each photodiode, are arranged. Incident wavelengths selected by the color filters enter the photodiodes, the output signals of the photodiodes are sequentially read out, and the spectral characteristics of the incident light are obtained by calculation processing. Figure 1b shows the cross-sectional view of a color sensor consisting of a metamaterial color filter formed on an Si photodiode through a spacer layer. The metamaterial color filter consists of a guided-mode resonant metal grating covered with a SiO_2 layer.

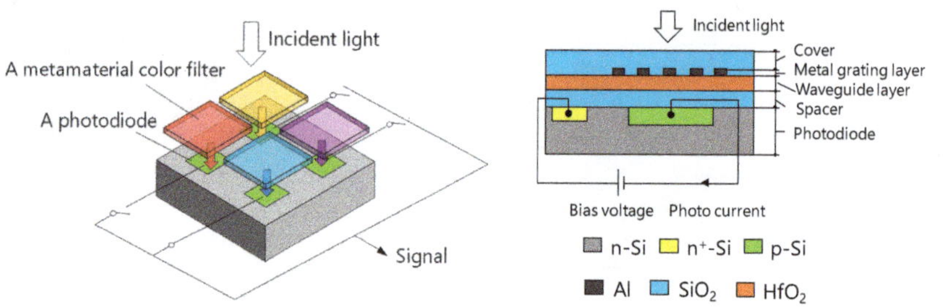

(a) (b)

Figure 1. Schematics of proposed devices. (**a**) Perspective view in the case of 2 × 2 color sensors; (**b**) cross-sectional view of one particular color sensor.

The guided-mode resonant metal grating consists of a 2D Al grating layer formed on an HfO_2 guided layer. The guided layer is formed on the SiO_2 spacer layer. Since the effective refractive index of the guided layer is higher than that of the surroundings, the guided layer functions as a planar waveguide and strongly confines the resonant wavelength. Transmitted light generates only zeroth order diffraction because the grating period is smaller than the resonant wavelength. The grating groove is filled with SiO_2. Incident light is impinged from the SiO_2 cover layer. Photocurrent generated in the photodiode is read out as an output signal. By changing the period of the metal grating layer of each color filter, it is possible to control the resonant wavelength and to extract spectral information from each photodiode.

3. Design and Numerical Analysis

3.1. Design of Color Filters

Numerical analysis was carried out using rigorous coupled-wave analysis (RCWA), which yields accurate results using Maxwell's equations in the frequency domain [37,38]. Figure 2 shows a calculation model of one color filter element. The electric field, magnetic field, and propagation direction of incident light are parallel to the *x*, *y*, and *z* axes, respectively. A 50-nm-thick SiO_2 cover layer, 30-nm-thick 2D-periodic Al nanodot array, 100-nm-thick HfO_2 waveguide layer, and a 150-nm-thick SiO_2 spacer are formed on an Si substrate of photodiode. Here, Λ and a are the grating period and nanodot size, respectively. The color filters have polarization independent because of 2D subwavelength gratings with the same periods for the *x* and *y* directions.

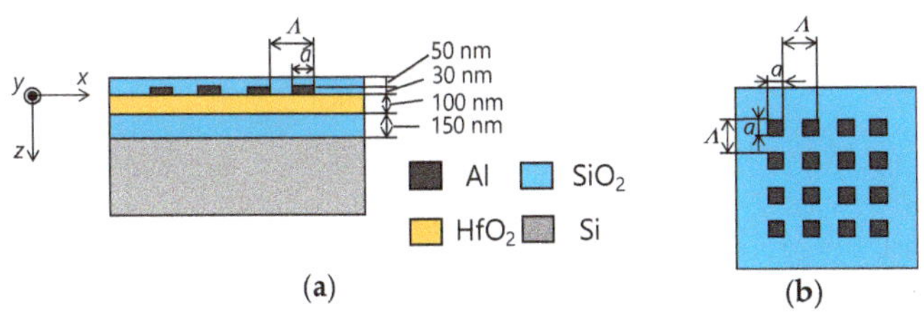

Figure 2. Calculation model of one color filter element. (**a**) Cross-sectional view; (**b**) top view.

Normally, incident light from outside was made to pass through the cover layer, and transmittance spectrum at the Si side of the SiO_2/Si interface was calculated to obtain spectral characteristics of the device. For optical designing of the structure integrated with the color filter and photodiode, spectral characteristics, including the influence of reflection and interference by the Si substrate, were calculated instead of just obtaining the transmission spectrum of only the color filter. The incident light polarized along the *x*-axis was normally incident and the total transmittance obtained by combining all the transmitted diffracted waves was calculated. Refractive indices of Al, Si, and SiO_2 in references [39–41], respectively, were used for the calculation. Figure 3 shows real and imaginary parts of the refractive index of HfO_2 used for the calculation, which was measured by a spectroscopic ellipsometer.

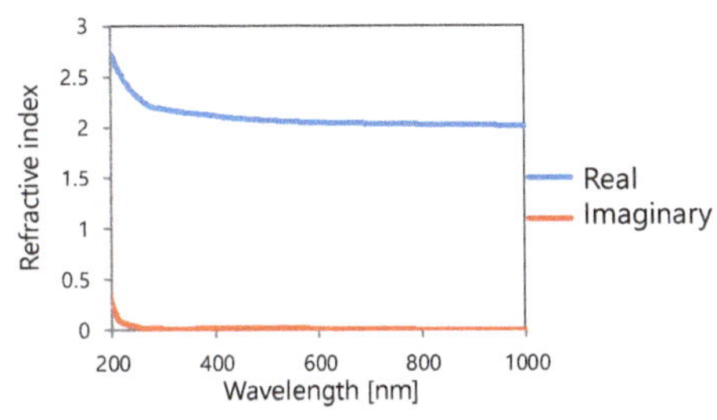

Figure 3. Refractive index of HfO_2 used for calculation.

Figure 4 shows the calculated transmittance spectra penetrating a Si substrate for several grating periods. Here, a/Λ was fixed to be 0.8. Figure 4 illustrates that the peak wavelength of the transmission spectrum can be controlled mainly through the grating period. The peak wavelength is shifted to the longer wavelength side as the period increases, which agrees with the principle of guided-mode resonant gratings. From this result, we could design the structures that exhibit functions of spectroscopes, including a reflection and interference effect, due to the influence of the interface between the SiO_2 space and Si photodiode. The output signal actually obtained by the color sensor is outputted as a characteristic signal, which is the product of the transmission spectrum shown in Figure 4 and the spectral sensitivity characteristic of the Si photodiode.

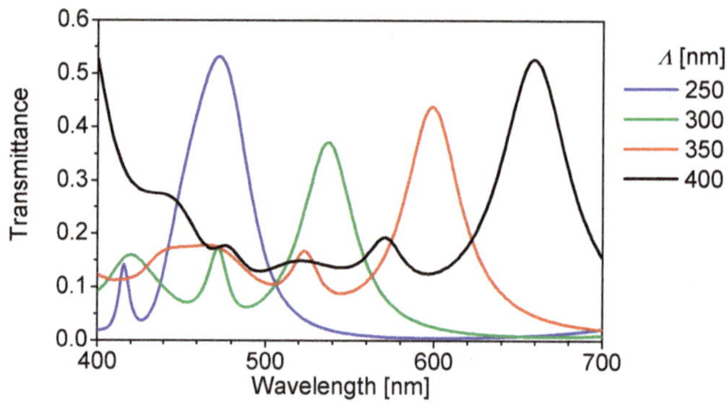

Figure 4. Calculated transmittance spectra penetrating a Si substrate for several grating periods. a/Λ is fixed to be 0.8.

Figure 5 shows the peak wavelength as a function of the grating period extracted from the results of Figure 4. It can be seen that the peak wavelength shifts in proportion to the grating period in the entire visible wavelength range. The linear fitting equation for Figure 5 is given by the following equation:

$$\lambda_{\text{peak}} = 1.24 \times \Lambda + 163.4 \tag{1}$$

Here, λ_{peak} is a peak wavelength.

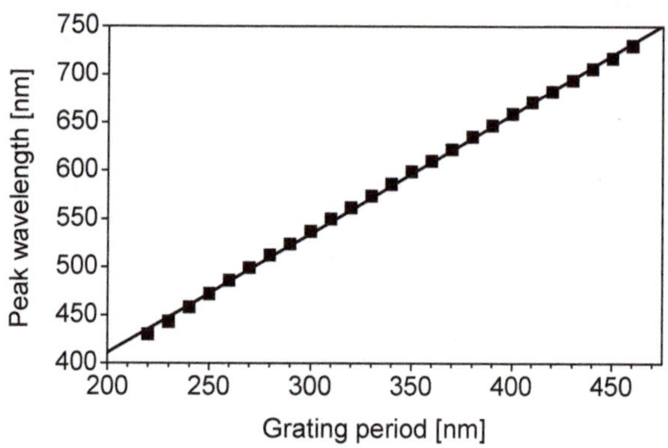

Figure 5. Peak wavelength as a function of grating period, extracted from the results of Figure 4.

3.2. Spectrum Reconstruction: Principle and Calculation Examples

An output signal from each color sensor is the integrated amount of the energy spectrum received by each photodiode and it cannot be decomposed into spectral components. Therefore, in order to reconstruct the spectral characteristic of the input light from the output signal, calculation between output signals is required. To solve the inverse problem of determining the input signals (spectrum of incident light) from the output signals, spectral characteristics are calculated using the Tikhonov regularization method [42]. Consider a spectroscopic device composed of b color sensors. b is the number of color sensors. The photocurrent of each color sensor is obtained as a product of the spectral characteristic of incident light and the wavelength sensitivity characteristic of each color sensor, integrated over the wavelength. The photocurrent is expressed in matrix format as follows:

$$\mathbf{O} = \mathbf{SI} \tag{2}$$

Here, $\mathbf{O}$ (A) is a $1 \times b$ column vector of photocurrent. $\mathbf{S}$ (A/W) is a $b \times c$ matrix of wavelength sensitivity, and $\mathbf{I}$ (W) is a $1 \times c$ column vector of spectral characteristics of incident light. c is a wavelength division number. $\mathbf{O}$ and $\mathbf{S}$ are measured experimentally, and $\mathbf{I}$ is to be determined. In order to solve this inverse problem, $\mathbf{I}$ is calculated using the matrix, $\mathbf{M}$, obtained by the Tikhonov regularization method, as shown below:

$$\mathbf{I} = \mathbf{MO} \tag{3}$$

$\mathbf{M}$ is a $c \times b$ matrix. Besides, because the spectral characteristic cannot be a negative value, the following condition is added:

$$\mathbf{I} = \mathbf{I}\ (\mathbf{I} > 0)\ \boldsymbol{or}\ 0\ (\mathbf{I} \leq 0) \tag{4}$$

Equations (3) and (4) are solved to obtain the spectral characteristics.

Next, several incident light spectra were designed (designed spectra) and compared with the incident light spectra (reconstructed spectra) calculated using Equations (3) and (4). It should be noted that b was set to 25 and the calculated wavelength range to be integrated was 400 to 700 nm. Twenty-five color filters with different grating periods between 210 and 450 nm at intervals of 10 nm were used for the matrix, $\mathbf{S}$. Some of the 25 filter characteristics are shown in Figure 4. Here, the sensitivity of the Si photodiode is ignored for simplicity.

The calculation results are shown in Figure 6. As shown in Figure 6a,b, the peak positions of the designed spectrum and the reconstructed spectrum coincide with each other, and the bandwidths also substantially coincide. However, the sharp edges of the designed spectrum take rounded shapes like the Gaussian distribution in the reconstructed spectrum. This can be attributed to the coarse resolution because just 25 filters are provided across a wavelength range of 300 nm. It can be improved by increasing the number of filters. Moreover, as can be seen from Figure 6c,d, it is possible to calculate the results for multiple peaks. Figure 6d reveals that as the width of the designed spectrum becomes narrower, the peak position of reconstructed data is in good agreement with the designed peak position, but the peak intensity of the reconstructed spectrum decreases. These results suggest that optical spectra in a wavelength range of 400 to 700 nm can be obtained by using the proposed devices; however, there is room for improvement.

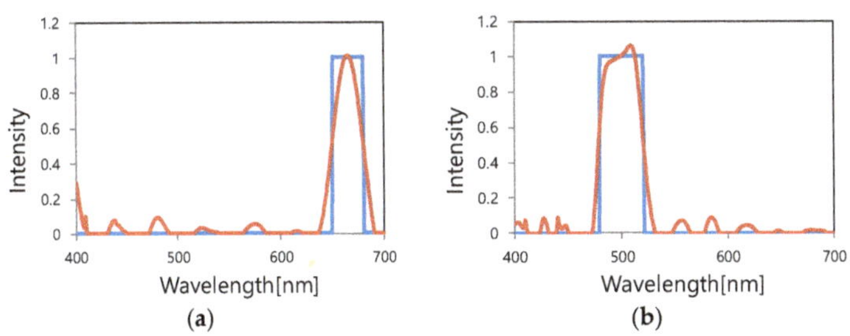

Figure 6. *Cont.*

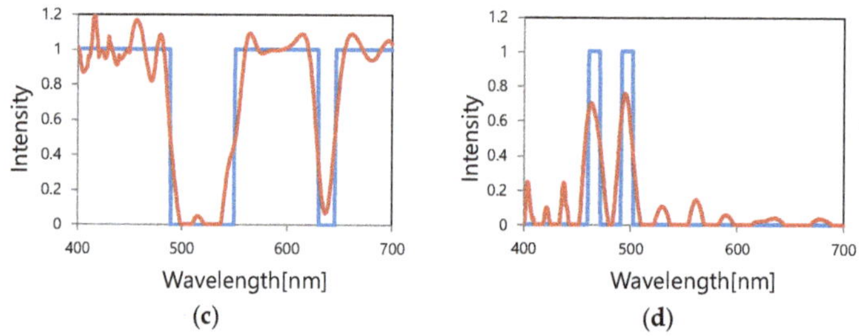

Figure 6. Simulation of spectral reconstruction (Blue line: Designed spectrum; red line: Reconstructed spectrum). (**a**) Single band spectrum near 700 nm in wavelength; (**b**) single band spectrum at a center wavelength of 500 nm; (**c**) multiple band-stop spectrum; (**d**) multiple peak spectrum.

4. Fabrication

Figure 7 shows the process steps. An n-type Si substrate of a 400 μm thickness is etched by a fast atom beam (FAB) to form alignment marks (Figure 7a,b). Next, to prevent metal contamination, an SiO_2 protective film with a thickness of 5 nm is formed by chemical vapor deposition (CVD) (Figure 7c). Then, P ions are implanted into the contact area of electrodes to form n+ -Si (Figure 7d). After that, ion implantation of B is performed to form p-Si (Figure 7e). Besides, rapid thermal annealing is performed to activate ions simultaneously with the recovery of crystal from damage due to ion implanted ions (Figure 7f). Next, as a spacer, SiO_2 is formed by CVD (Figure 7g). Then, as a waveguide layer, HfO_2 is deposited by electron-beam (EB) evaporation (Figure 7h).

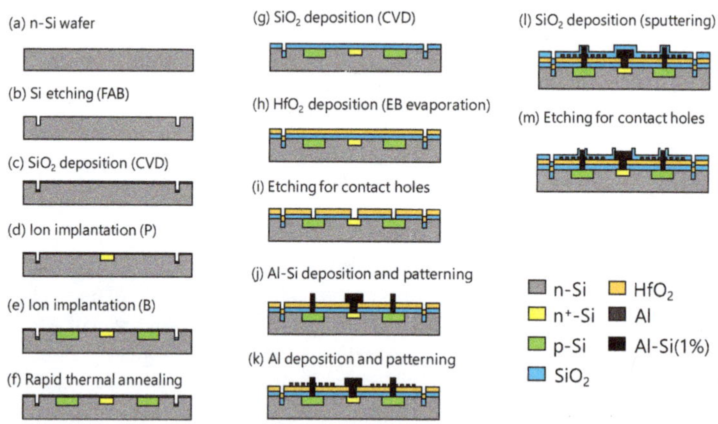

Figure 7. Process steps.

After that, SiO_2 and HfO_2 etching is performed to make contact holes (Figure 7i). Next, the Al-Si (1%) layer with a thickness of 400 nm is formed by sputtering, and wet etching is performed to form electrodes (Figure 7j). After Al film formation by sputtering, nanodot array structures are patterned by EB lithography, followed by FAB etching (Figure 7k). Next, an SiO_2 cover layer is formed by sputtering (Figure 7l). Finally, SiO_2 in the electrode pad portion is etched (Figure 7m).

Figure 8 shows a reflection image of an optical microscope of color filters fabricated on a photodiode array. The filters are fabricated in 25 patterns, with grating periods between 220 and 460 nm at increments of 10 nm, which correspond (a) to (y) in the figure. All filters are 150 μm × 150 μm in size. Although the color filter is designed as a transmissive type, the structural color, which depends on the period of the structure, can also be confirmed in the reflected images. The wiring pattern connected from each photodiode can also be confirmed.

Figure 9 shows scanning electron microscope (SEM) images of the fabricated color filters corresponding to the symbols (a) to (y) of Figure 8. All filters are designed to be 0.8 of the a/Λ ratio. It can be seen that all the filters are fabricated accurately.

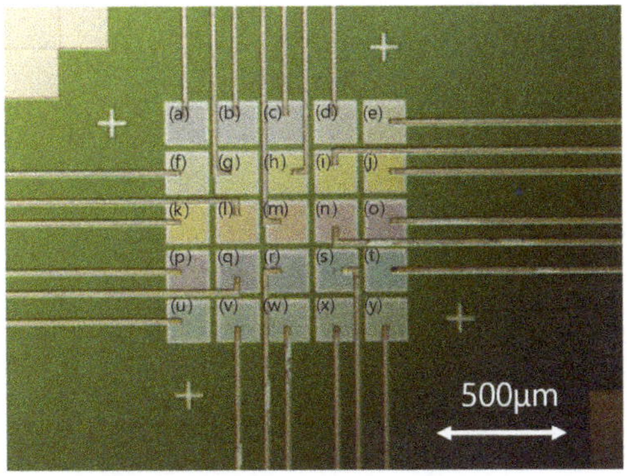

Figure 8. Optical microphotographs of fabricated color filters on the photodiode array. (**a**) Λ = 220 nm, (**b**) Λ = 230 nm, (**c**) Λ = 240 nm, (**d**) Λ = 250 nm, (**e**) Λ = 260 nm, (**f**) Λ = 270 nm, (**g**) Λ = 280 nm, (**h**) Λ = 290 nm, (**i**) Λ = 300 nm, (**j**) Λ = 310 nm, (**k**) Λ = 320 nm, (**l**) Λ = 330 nm, (**m**) Λ = 340 nm, (**n**) Λ = 350 nm, (**o**) Λ = 360 nm, (**p**) Λ = 370 nm, (**q**) Λ = 380 nm, (**r**) Λ = 390 nm, (**s**) Λ = 400 nm, (**t**) Λ = 410 nm, (**u**) Λ = 420 nm, (**v**) Λ = 430 nm, (**w**) Λ = 440 nm, (**x**) Λ = 450 nm, and (**y**) Λ = 460 nm.

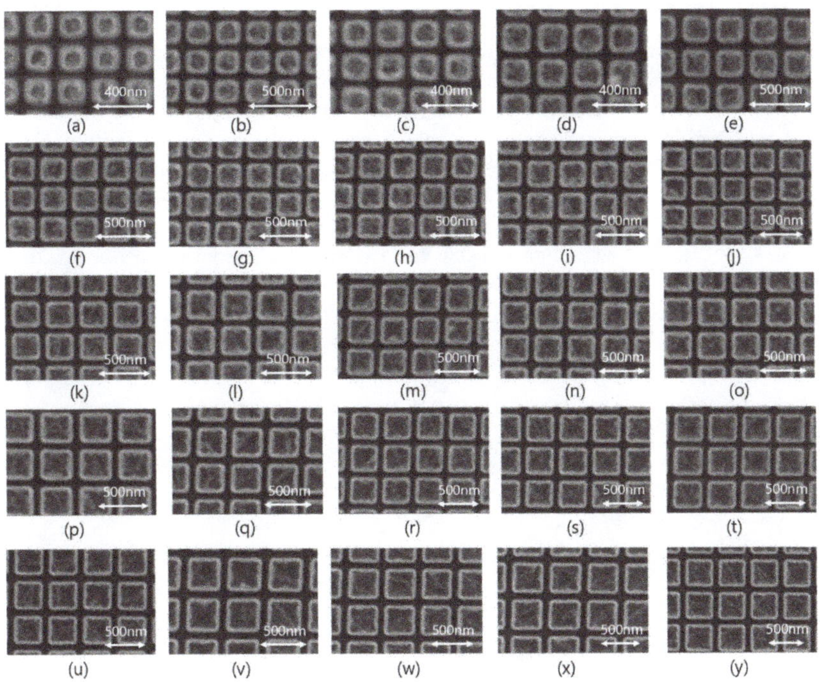

Figure 9. SEM images of fabricated color filters. (**a**) Λ = 220 nm, (**b**) Λ = 230 nm, (**c**) Λ = 240 nm, (**d**) Λ = 250 nm, (**e**) Λ = 260 nm, (**f**) Λ = 270 nm, (**g**) Λ = 280 nm, (**h**) Λ = 290 nm, (**i**) Λ = 300 nm, (**j**) Λ = 310 nm, (**k**) Λ = 320 nm, (**l**) Λ = 330 nm, (**m**) Λ = 340 nm, (**n**) Λ = 350 nm, (**o**) Λ = 360 nm, (**p**) Λ = 370 nm, (**q**) Λ = 380 nm, (**r**) Λ = 390 nm, (**s**) Λ = 400 nm, (**t**) Λ = 410 nm, (**u**) Λ = 420 nm, (**v**) Λ = 430 nm, (**w**) Λ = 440 nm, (**x**) Λ = 450 nm, and (**y**) Λ = 460 nm.

5. Measured Results and Discussion

Spectral sensitivity characteristics of color sensors were measured using a broadband spectral response measurement system (CEP-25BXS, Bunkou Keiki Co., Ltd., Tokyo, Japan). In this measurement, the wavelength resolution was set to 10 nm. Different color sensors exhibited different spectral sensitivity characteristics. Figure 10 shows the relationship between the grating period and the peak wavelength of the spectral sensitivity. The peak wavelength is linearly dependent on the grating period, and the slope of the straight line almost agrees with that of the calculated peak wavelength shift of the color filters shown in Figure 5. The linear fitting equation for Figure 10 is given by the following equation:

$$\lambda_{\text{peak}} = 1.28 \times \Lambda + 105.4 \quad (5)$$

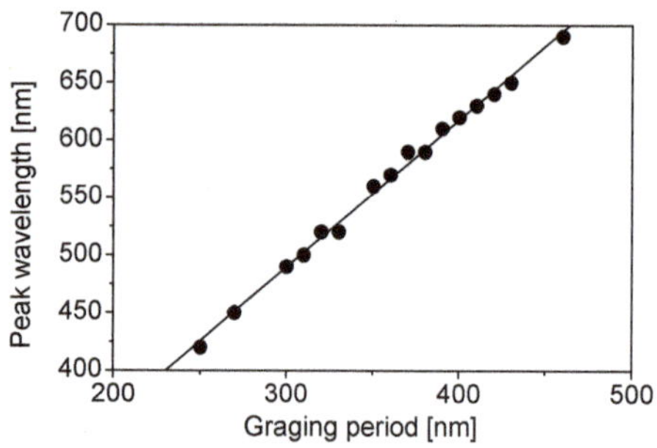

Figure 10. Dependence of the peak wavelength of the spectral sensitivity on the grating period.

When the monochromatic light with wavelengths of 450, 500, 550, 600, and 650 nm is incident, the spectral characteristics of the incident light are reconstructed from the output signals obtained by the respective color sensors. Spectral characteristics of the reconstructed incident light are shown with solid lines in Figure 11. For comparison, spectral characteristics of the incident light measured with a commercially available spectrometer (HR4000CG-UV-NIR, Ocean optics, Inc., Largo, FL, USA) are shown as original spectra with dotted lines. The peak wavelength of the reconstructed spectrum can be confirmed to be near that of the original spectrum. Therefore, we believe that the spectroscopic measurement was successfully performed using the fabricated device. However, reconstructed spectra become broader compared to the original spectra. This increment in width can be attributed to the coarse resolution because only 25 filters are provided across a wavelength range of 300 nm. The reconstructed spectra can be improved by increasing the number of filters.

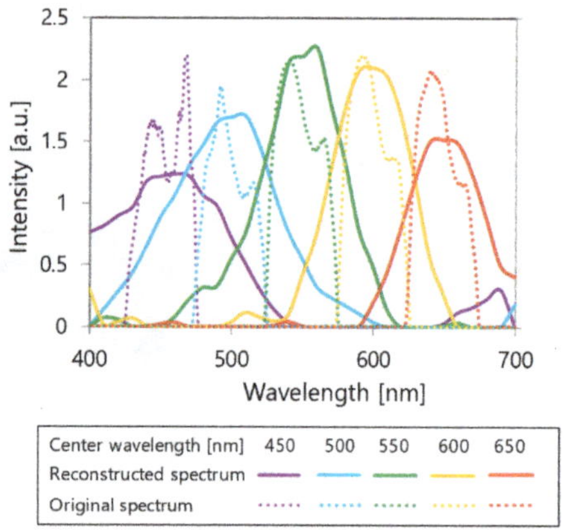

Figure 11. Spectral reconstruction.

Figure 12 shows the relationship between the peak wavelengths of reconstructed spectra and the center wavelength of monochromatic lights extracted from the results of Figure 11. It is found that the peak wavelengths of the reconstructed spectra are in good agreement with the center wavelengths of the monochromatic lights.

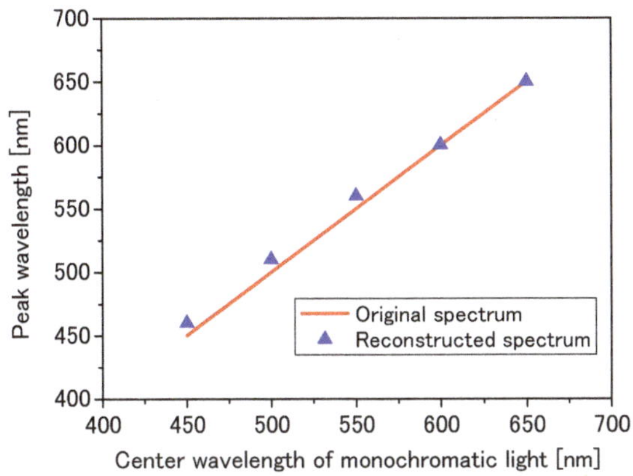

Figure 12. Dependence of the peak wavelengths of reconstructed spectra on the center wavelength of the incident monochromatic lights.

6. Conclusions

We fabricated a filter array spectroscope consisting of metamaterial color filters and Si photodiodes. The color filters based on guided-mode resonant metal gratings were newly designed. The peak wavelength of the transmission spectrum could be controlled mainly by the grating period. The filters were fabricated in 25 patterns, with grating periods between 220 and 460 nm at increments of 10 nm. SEM images of fabricated color filters confirmed that all the filters were fabricated accurately. The peak wavelength of the spectral sensitivity was found to be linearly dependent on the grating period. Several monochromatic lights were made incident on the fabricated device and the spectral characteristics of the incident light were reconstructed from the output signals obtained from the respective color sensors. The peak wavelengths of the reconstructed spectra were in good agreement with the center wavelengths of the monochromatic lights.

Author Contributions: Conceptualization, Y.K.; methodology, Y.K.; software, D.E.; validation, Y.K., D.E. and K.H.; formal analysis, Y.K. and D.E.; investigation, Y.K., D.E. and K.H.; resources, Y.K., D.E. and K.H.; data curation, D.E. and Y.K.; writing—original draft preparation, Y.K.; writing—review and editing, Y.K.; visualization, D.E. and Y.K.; supervision, K.H.; project administration, Y.K.; funding acquisition, Y.K.

Funding: A part of this research was funded by the MEXT KAKENHI 16K13648 and 16H04342.

Acknowledgments: Authors thank H. Sai for providing technical support for spectral response measurement. A part of this research was performed in Micro/Nano-Machining Research and Education Center (Tohoku University) and the Center for Integrated Nano Technology Support (Tohoku University) with assistance of MEXT Nanotechnology Platform.

Conflicts of Interest: The authors declare no conflict of interest.

References

1. Ebbesen, T.W.; Lezec, H.J.; Ghaemi, H.F.; Thio, T.; Wolff, P.A. Extraordinary optical transmission through sub-wavelength hole arrays. *Nature* **1998**, *391*, 667–669. [CrossRef]
2. Genet, C.; Ebbesen, T.W. Light in tiny holes. *Nature* **2007**, *445*, 39–46. [CrossRef] [PubMed]
3. Kaplan, A.F.; Xu, T.; Guo, L.J. High efficiency resonance-based spectrum filters with tunable transmission bandwidth fabricated using nanoimprint lithography. *Appl. Phys. Lett.* **2011**, *99*, 143111. [CrossRef]

4. Xu, T.; Wu, Y.K.; Luo, X.; Guo, L.J. Plasmonic nanoresonators for high-resolution colour filtering and spectral imaging. *Nat. Commun.* **2010**, *1*, 59. [CrossRef] [PubMed]
5. Lee, H.S.; Yoon, Y.T.; Lee, S.S.; Kim, S.H.; Lee, K.D. Color filter based on a subwavelength patterned metal grating. *Opt. Express* **2007**, *15*, 15457–15463. [CrossRef] [PubMed]
6. Chen, Q.; Cumming, D.R.S. High transmission and low color cross-talk plasmonic color filters using triangular-lattice hole arrays in aluminum films. *Opt. Express* **2010**, *18*, 14056–14062. [CrossRef] [PubMed]
7. Chou, S.Y.; Krauss, P.R.; Renstrom, P.J. Nanoimprint lithography. *J. Vac. Sci. Technol. B* **1996**, *14*, 4129–4133. [CrossRef]
8. Chou, S.Y.; Krauss, P.R.; Zhang, W.; Guo, L.; Zhuang, L. Sub-10 nm imprint lithography and applications. *J. Vac. Sci. Technol. B* **1997**, *15*, 2897–2904. [CrossRef]
9. Bao, L.R.; Cheng, X.; Huang, X.D.; Guo, L.J.; Pang, S.W.; Yee, A.F. Nanoimprinting over topography and multilayer three-dimensional printing. *J. Vac. Sci. Technol. B* **2002**, *20*, 2881–2886. [CrossRef]
10. Zhang, W.; Chou, S.Y. Fabrication of 60-nm transistors on 4-in. wafer using nanoimprint at all lithography levels. *Appl. Phys. Lett.* **2003**, *83*, 1632–1634. [CrossRef]
11. Alkaisi, M.M.; Jayatissa, W.; Konijn, M. Multilevel nanoimprint lithography. *Curr. Appl. Phys.* **2004**, *4*, 111–114. [CrossRef]
12. Roy, E.; Kanamori, Y.; Belotti, M.; Chen, Y. Enhanced UV imprint ability with a tri-layer stamp configuration. *Microelectron. Eng.* **2005**, *78–79*, 689–694. [CrossRef]
13. Kanamori, Y.; Okochi, M.; Hane, K. Fabrication of antireflection subwavelength gratings at the tips of optical fibers using UV nanoimprint lithography. *Opt. Express* **2013**, *21*, 322–328. [CrossRef] [PubMed]
14. Ito, S.; Kikuchi, E.; Watanabe, M.; Sugiyama, Y.; Kanamori, Y.; Nakagawa, M. Silica imprint templates with concave patterns from single-digit nanometers fabricated by electron beam lithography involving argon ion beam milling. *Jpn. J. Appl. Phys.* **2017**, *56*, 06GL01. [CrossRef]
15. Lee, S.W.; Lee, K.S.; Ahn, J.; Lee, J.J.; Kim, M.G.; Shin, Y.B. Highly sensitive biosensing using arrays of plasmonic Au nanodisks realized by nanoimprint lithography. *ACS Nano* **2011**, *5*, 897–904. [CrossRef] [PubMed]
16. Peer, A.; Biswas, R. Extraordinary optical transmission in nanopatterned ultrathin metal films without holes. *Nanoscale* **2016**, *8*, 4657–4666. [CrossRef] [PubMed]
17. Nishiguchi, K.; Sueyoshi, K.; Hisamoto, H.; Endo, T. Fabrication of gold-deposited plasmonic crystal based on nanoimprint lithography for label-free biosensing application. *Jpn. J. Appl. Phys.* **2016**, *55*, 08RE02. [CrossRef]
18. Peer, A.; Hu, Z.; Singh, A.; Hollingsworth, J.A.; Biswas, R.; Htoon, H. Photoluminescence enhancement of $CuInS_2$ quantum dots in solution coupled to plasmonic gold nanocup array. *Small* **2017**, *13*, 1700660. [CrossRef] [PubMed]
19. Hutley, M.C. Diffraction gratings. In *Diffractive Optics for Industrial and Commercial Applications*; Turunen, J., Wyrowski, F., Eds.; Akademie Verlag: Berlin, Germany, 1997; pp. 59–80. ISBN 3055017331.
20. Ema, D.; Kanamori, Y.; Sai, H.; Hane, K. Plasmonic color filters integrated on a photodiode array. *Electron. Commun. Jpn.* **2018**, *101*, 95–104. [CrossRef]
21. Chang, C.C.; Lin, N.T.; Kurokawa, U.; Choi, B.I. Spectrum reconstruction for filter-array spectrum sensor from sparse template selection. *Opt. Eng.* **2011**, *50*, 114402. [CrossRef]
22. Kurokawa, U.; Choi, B.I.; Chang, C.C. Filter-Based Miniature Spectrometers: Spectrum Reconstruction Using Adaptive Regularization. *IEEE Sens. J.* **2011**, *11*, 1556–1563. [CrossRef]
23. Chang, C.C.; Lin, H.Y. Spectrum Reconstruction for On-Chip Spectrum Sensor Array Using a Novel Blind Nonuniformity Correction Method. *IEEE Sens. J.* **2012**, *12*, 2586–2592. [CrossRef]
24. Chang, C.C.; Su, Y.J.; Kurokawa, U.; Choi, B.I. Interference Rejection Using Filter-Based Sensor Array in VLC Systems. *IEEE Sens. J.* **2012**, *12*, 1025–1032. [CrossRef]
25. Oliver, J.; Lee, W.B.; Lee, H.N. Filters with random transmittance for improving resolution in filter-array-based spectrometers. *Opt. Express* **2013**, *21*, 3969–3989. [CrossRef] [PubMed]
26. Oliver, J.; Lee, W.; Park, S.; Lee, H.N. Improving resolution of miniature spectrometers by exploiting sparse nature of signals. *Opt. Express* **2012**, *20*, 2613–2625. [CrossRef] [PubMed]
27. Norton, S.M.; Erdogan, T.; Morris, G.M. Coupled-mode theory of resonant-grating filters. *J. Opt. Soc. Am. A* **1997**, *14*, 629–639. [CrossRef]

28. Jacob, D.K.; Dunn, S.C.; Moharam, M.G. Design considerations for narrow-band dielectric resonant grating reflection filters of finite length. *J. Opt. Soc. Am. A* **2000**, *17*, 1241–1249. [CrossRef]
29. Brundrett, D.L.; Glytsis, E.N.; Gaylord, T.K.; Bendickson, J.M. Effects of modulation strength in guided-mode resonant subwavelength gratings at normal incidence. *J. Opt. Soc. Am. A* **2000**, *17*, 1221–1230. [CrossRef]
30. Peters, D.W.; Kemme, S.A.; Hadley, G.R. Effect of finite grating, waveguide width, and end-facet geometry on resonant subwavelength grating reflectivity. *J. Opt. Soc. Am. A* **2004**, *21*, 981–987. [CrossRef]
31. Kobayashi, T.; Kanamori, Y.; Hane, K. Surface laser emission from solid polymer dye in a guided mode resonant grating filter structure. *Appl. Phys. Lett.* **2005**, *87*, 151106. [CrossRef]
32. Kanamori, Y.; Kitani, T.; Hane, K. Guided-Mode Resonant Grating Filter Fabricated on Silicon-on-Insulator Substrate. *Jpn. J. Appl. Phys.* **2006**, *45*, 1883–1885. [CrossRef]
33. Kanamori, Y.; Kitani, T.; Hane, K. Control of guided resonance in a photonic crystal slab using microelectromechanical actuators. *Appl. Phys. Lett.* **2007**, *90*, 031911. [CrossRef]
34. Kanamori, Y.; Katsube, H.; Furuta, T.; Hasegawa, S.; Hane, K. Design and Fabrication of Structural Color Filters with Polymer-Based Guided-Mode Resonant Gratings by Nanoimprint Lithography. *Jpn. J. Appl. Phys.* **2009**, *48*, 06FH04. [CrossRef]
35. Kanamori, Y.; Ozaki, T.; Hane, K. Fabrication of Ultrathin Color Filters for Three Primary Colors Using Guided-Mode Resonance in Silicon Subwavelength Gratings. *Opt. Rev.* **2014**, *21*, 723–727. [CrossRef]
36. Kanamori, Y.; Ozaki, T.; Hane, K. Reflection color filters of the three primary colors with wide viewing angles using common-thickness silicon subwavelength gratings. *Opt. Express* **2014**, *22*, 25663–25672. [CrossRef] [PubMed]
37. Moharam, M.G.; Grann, E.B.; Pommet, D.A.; Gaylord, T.K. Formulation for stable and efficient implementation of the rigorous coupled-wave analysis of binary gratings. *J. Opt. Soc. Am. A* **1995**, *12*, 1068–1076. [CrossRef]
38. Lalanne, P.; Morris, G.M. Highly improved convergence of the coupled-wave method for TM polarization. *J. Opt. Soc. Am. A* **1996**, *13*, 779–784. [CrossRef]
39. Smith, D.Y.; Shiles, E.; Inokuti, M. The Optical Properties of Metallic Aluminum. In *Handbook of Optical Constants of Solids*; Palik, E.D., Ed.; Academic Press: Orlando, FL, USA, 1985; Volume 1, pp. 369–406. ISBN 0125444206.
40. Edwards, D.F. Silicon (Si). In *Handbook of Optical Constants of Solids*; Palik, E.D., Ed.; Academic Press: Orlando, FL, USA, 1985; Volume 1, pp. 547–569. ISBN 0125444206.
41. Philipp, H.R. Silicon Dioxide (SiO2) (Glass). In *Handbook of Optical Constants of Solids*; Palik, E.D., Ed.; Academic Press: Orlando, FL, USA, 1985; Volume 1, pp. 749–763. ISBN 0125444206.
42. Aster, R.C.; Borchers, B.; Thurber, C.H. *Parameter Estimation and Inverse Problems*, 2nd ed.; Academic Press: Waltham, MA, USA, 2012; pp. 93–127. ISBN 9780128100929.

materials

MDPI

Article

Elimination of Unwanted Modes in Wavelength-Selective Uncooled Infrared Sensors with Plasmonic Metamaterial Absorbers using a Subtraction Operation

Shinpei Ogawa [1,*], Yousuke Takagawa [2] and Masafumi Kimata [2]

[1] Advanced Technology R&D Center, Mitsubishi Electric Corporation, 8-1-1 Tsukaguchi-Honmachi, Amagasaki, Hyogo 661-8661, Japan

[2] College of Science and Engineering, Ritsumeikan University, 1-1-1 Noji-higashi, Kusatsu, Shiga 525-8577, Japan; Takagawa.Yosuke@ds.MitsubishiElectric.co.jp (Y.T.); kimata@se.ritsumei.ac.jp (M.K.)

* Correspondence: Ogawa.Shimpei@eb.MitsubishiElectric.co.jp; Tel.: +81-6-6497-7533

Received: 3 September 2019; Accepted: 25 September 2019; Published: 27 September 2019

Abstract: Wavelength- or polarization-selective uncooled infrared (IR) sensors have various applications, such as in fire detection, gas analysis, hazardous material recognition, biological analysis, and polarimetric imaging. The unwanted modes originating due to the absorption by the materials used in these sensors, other than plasmonic metamaterial absorbers (PMAs), cause serious issues by degenerating the wavelength or polarization selectivity. In this study, we demonstrate a method for eliminating these unwanted modes in wavelength- or polarization-selective uncooled IR sensors with various PMAs, using a subtraction operation and a reference pixel. The aforementioned sensors and the reference pixels were fabricated using a complementary metal oxide semiconductor and micromachining techniques. We fabricated the reference pixel with the same structure as the PMA sensors, except a flat mirror was formed on the absorber surface instead of PMAs. The spectral responsivity measurements demonstrated that single-mode detection can be achieved through the subtraction operation with the reference pixel. The method demonstrated in this study can be applied to any type of uncooled IR sensors to create high-performance wavelength- or polarization-selective absorbers capable of multispectral or polarimetric detection.

Keywords: plasmonics; metamaterials; uncooled; IR sensors; wavelength-selective

1. Introduction

Uncooled infrared (IR) sensors with micro-electro-mechanical system (MEMS) based pixel structures are used in a wide range of applications, such as in security, surveillance, maintenance, firefighting, and in the automotive industry [1,2]. Recently, there has been increased interest in developing advanced functional uncooled IR sensors with wavelength- or polarization-selectivity, especially due to their applicability in fire detection, gas analysis, hazardous material recognition, biological analysis, and polarimetric imaging [3,4]. The wavelength-selective function is used for analyzing and recognizing objects by spectral information, and the polarization-selective function can contribute to enhancing object recognition, such as face recognition using polarimetric information [5]. We have previously demonstrated wavelength- or polarization-selective uncooled IR sensors using various plasmonic metamaterial absorbers (PMAs), such as plasmonic crystal (PC) [6–9], metal-insulator-metal (MIM) [3,10], and mushroom-type PMAs [11–13] at the middle-wavelength IR (MWIR) and the long-wavelength IR (LWIR) regions. The PMAs have periodic metal surface structures, which support the induced surface plasmon modes and lead to the wavelength- or polarization-selective absorption. Therefore, the absorption wavelength and the polarization can be structurally controlled by the surface patterns of these structures. Such PMAs can realize wavelength- or polarization-selective functions in uncooled IR sensors without filters or polarizers, which lead to low-cost fabrication and enable different pixels to be integrated in an array. Moreover, PMAs with smaller and thinner absorbers have significant advantages over traditional absorbers.

Several researchers have studied the effects of various surface patterns of PMAs to realize single [14–17], multi-mode [18–23], and broadband absorption [24–28] in broad-wavelength regions ranging from ultraviolet [29] to terahertz [30,31]. However, only a few investigations have been performed that demonstrate the application of PMAs in actual devices such as uncooled IR sensors [32]. In some cases, unwanted absorption modes have been observed, which cause additional absorption of wavelengths, and therefore degenerated performance with regard to wavelength or polarization selectivity. These unwanted modes cannot be attributed to the propagating or localized surface plasmon resonance induced by the surface patterns of the PMAs. It is extremely important to eliminate such unwanted absorption modes to realize high-performance wavelength- or polarization-selective uncooled IR sensors. In this study, we investigate the origin of unwanted absorption modes and develop a method to eliminate such absorption modes.

2. Elimination of Unwanted Modes

2.1. Origin of Unwanted Modes

Figure 1a and b shows schematic illustrations of a typical thermopile sensor with the cross-section of PMAs as uncooled IR sensors and the side view of the PMA sensor with unwanted IR absorptions, respectively. In this study, we adopted two-dimensional (2D) PC-type PMAs (2D PC-PMAs) with a thermopile [33] as a MEMS-based uncooled IR sensor. A 2D PC-PMA produces wavelength-selective absorption at a wavelength nearly equal to its surface period [6,7].

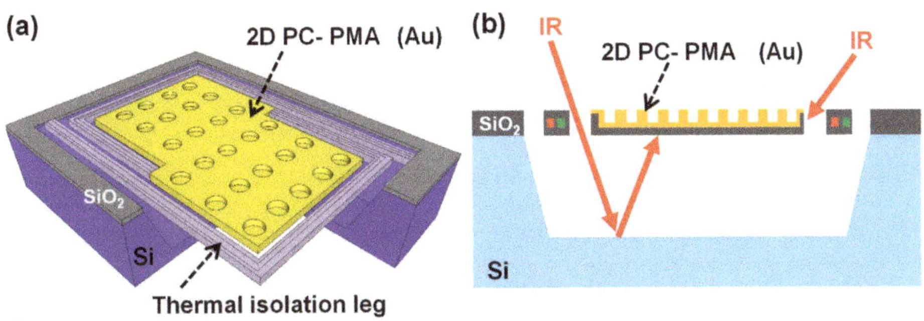

Figure 1. Schematic illustrations of (**a**) typical micro-electro-mechanical system (MEMS) based thermopile with 2D PC-PMA and (**b**) unwanted absorption at the backside and the sides of the 2D PC-PMA.

We had previously demonstrated that the unwanted absorption occurs owing to the absorptions at the backside and other sides of SiO_2 in a 2D PC-PMA (Figure 1b). To address this issue, an Al reflection layer was inserted at the backside and the sides of the 2D PC-PMAs, which were coated with Au. This drastically reduced unwanted absorption in the system [7]. Additionally, clear wavelength or polarization selectivity was achieved by restricting the incident IR ray to only the absorber area using a pinhole. Without the pinhole, we could still observe unwanted absorption in the LWIR region.

A schematic of the thermopile with the flat mirror surface (reference pixel) prepared in this study is shown in Figure 2a. Its responsivity and absorbance were measured to investigate the origin of these unwanted modes. Figure 2b shows the measured spectral responsivity of the reference pixel and the calculated value of absorption by the 1.5 μm thick SiO_2, which corresponds to the actual thickness of the sensors, as shown in Figure 2a. The details of the fabrication procedure and the measurement systems are explained in subsequent sections. As expected, the reference pixel produced no output signal because the surface mirror and the backside reflector reflected all the incident IR rays, and there was no absorption of IR rays. The calculations were performed using the rigorous coupled wave analysis (RCWA) method [34]. RSOFT DiffractMOD software (version 2018.12-1, Synopsys, Inc., California, CA, USA) was used for the RCWA calculation. Unexpectedly, responsivity peaks were observed in the LWIR region, which corresponded to the calculated absorbance of SiO_2. The calculated absorbance spectra were attributed to the thin-film interference in the thickness direction of SiO_2. These results demonstrate that the origin of the unwanted modes can be attributed to the intrinsic absorption of SiO_2 used in the sensor structures as the thermal isolation legs (Figure 2c).

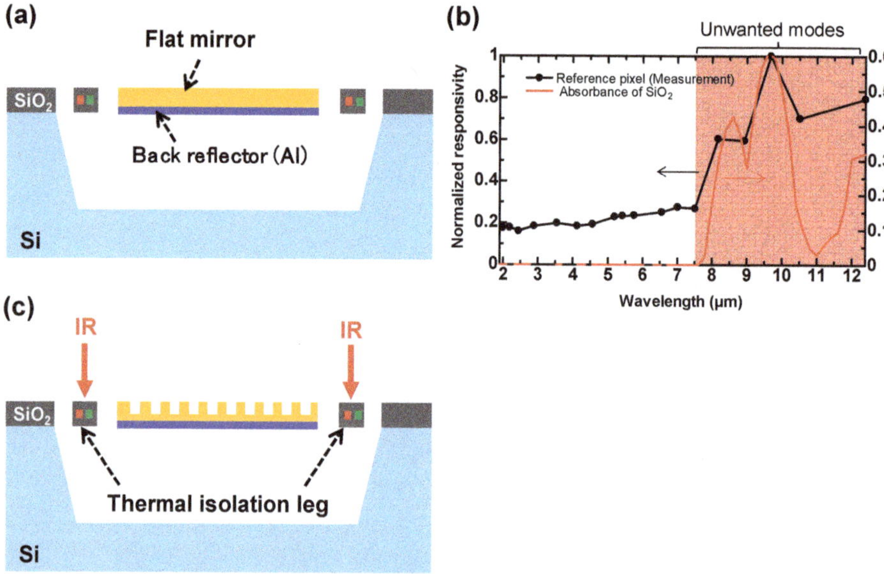

Figure 2. (**a**) Schematic illustration of a typical MEMS-based thermopile with flat mirror (reference pixel) and (**b**) its measured spectral responsivity and calculated absorbance of 1.5 μm thick SiO_2. (**c**) Schematic illustration of unwanted absorption by the SiO_2-thermal isolation legs in typical MEMS-based uncooled IR sensors.

The thermal isolation legs are typically used in most MEMS-based uncooled IR sensor structures. Therefore, the unwanted absorption by the thermal isolation legs is a common obstacle to realizing high performance in wavelength- or polarization-selective uncooled IR sensors. It is extremely difficult to block the top surface, the sides, and the backside of the thermal isolation legs to avoid any unwanted absorption in these sensors, and therefore we present a subtraction operation for eliminating the unwanted modes.

2.2. Subtraction Operation Using Reference Pixel

Figure 3a shows the proposed concept to eliminate unwanted modes using a subtraction operation with the reference pixel. The sensor with PMAs has a wavelength-selective detection peak caused by the plasmonic resonance in PMAs and relatively broad detection in the LWIR region owing to the absorption by the SiO_2 present in the thermal isolation legs. The reference pixel was designed with the same sensor structure except for the absorber surface in which a flat mirror was used as the absorber surface instead of PMAs. Therefore, the reference pixel shows the same unwanted absorption modes as the sensor with PMAs, which is caused by the SiO_2-thermal isolation legs. Figure 3b shows the configuration of the sensor with PMA and the reference pixel with a flat mirror for the subtraction operation. The output signal between Pad 1 and Pad 2 was measured, whereby the subtraction voltage ($V_S - V_R$) between the output voltage of the sensor with the PMA (V_S) and that of the reference pixel (V_R) was obtained.

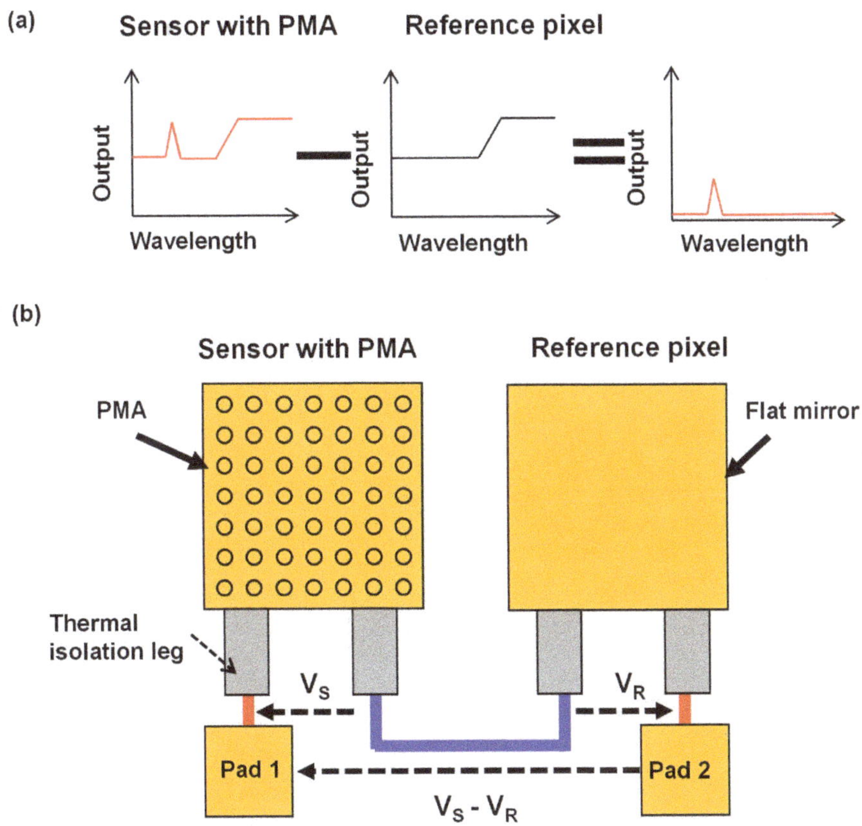

Figure 3. (**a**) Demonstration of the concept to eliminate the unwanted modes using a subtraction operation with a reference pixel. (**b**) Configuration for the subtraction operation between a thermopile with PMA and a reference pixel with flat mirror.

3. Sensor Fabrication

Figure 4 shows the actual configuration of the sensor with 2D PC-PMAs and the reference pixel. The signal lines to measure voltages were connected as shown in Figure 3b.

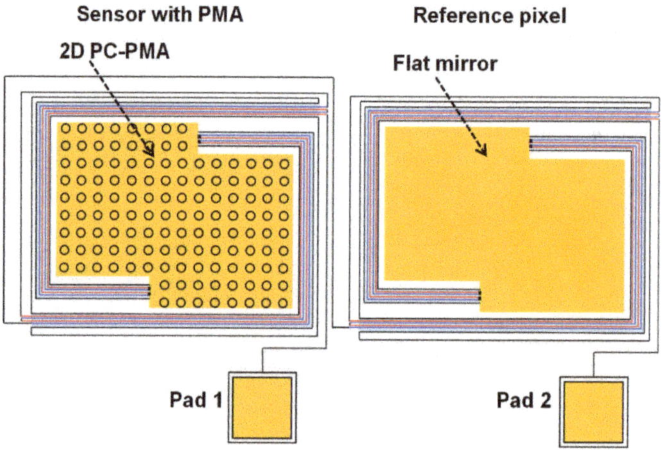

Figure 4. Schematic illustration of the sensor with 2D PC-PMA and the reference pixel, in which the output signal was produced from the subtraction operation between them.

Figure 5a shows the fabrication procedure of the sensors with 2D PC-PMAs and the reference pixel with the subtraction operation. The sensors with 2D PC-PMAs and the reference pixels were fabricated on a six inch silicon (Si) substrate using a standard complementary metal oxide semiconductor (CMOS) process, the details of which have been reported in our previous studies [7–9,20]. A series of p- and n-type poly-Si regions formed thermocouples, and a selected amount of ion implantation controlled their resistivity. An Al layer was deposited under the absorber area as a backside reflective layer, and holes for cavities were formed via reactive-ion etching (RIE). Subsequently, a 1.5 μm thick SiO_2 layer was deposited on the absorber area (Figure 5a, i). Following this, the 2D PC-PMA structures were fabricated over the SiO_2 layer of the IR absorber area of the sensors by using the RIE process (Figure 5a, ii), whereas no pattern was formed in the reference pixels. Next, a 50 nm thick Cr adhesive layer and 250 nm thick Au surface plasmon resonance layer were deposited by sputtering on the entire surface of the wafer. The Au layer was sufficiently thicker than the skin depth of the IR wavelengths, therefore, absorption of the incident IR rays by the Cr and SiO_2 layers beneath the Au layer could not occur. The Cr and Au layers were selectively etched using a wet etchant, except the part of the layers covering the 2D PC-PMA and flat mirror regions. (Figure 5a, iii). We confirmed that the Cr and Au layers were uniformly coated using scanning electron microscopy observation. Each wafer was then diced into chips. The Si of every chip was anisotropically etched through the holes using tetramethylammonium hydroxide (TMAH), as shown in Figure 5a, iv. The backside reflective Al layer was not etched because TMAH was doped with Si. The thermally isolated freestanding structure was completed in the cavity under the IR absorber area.

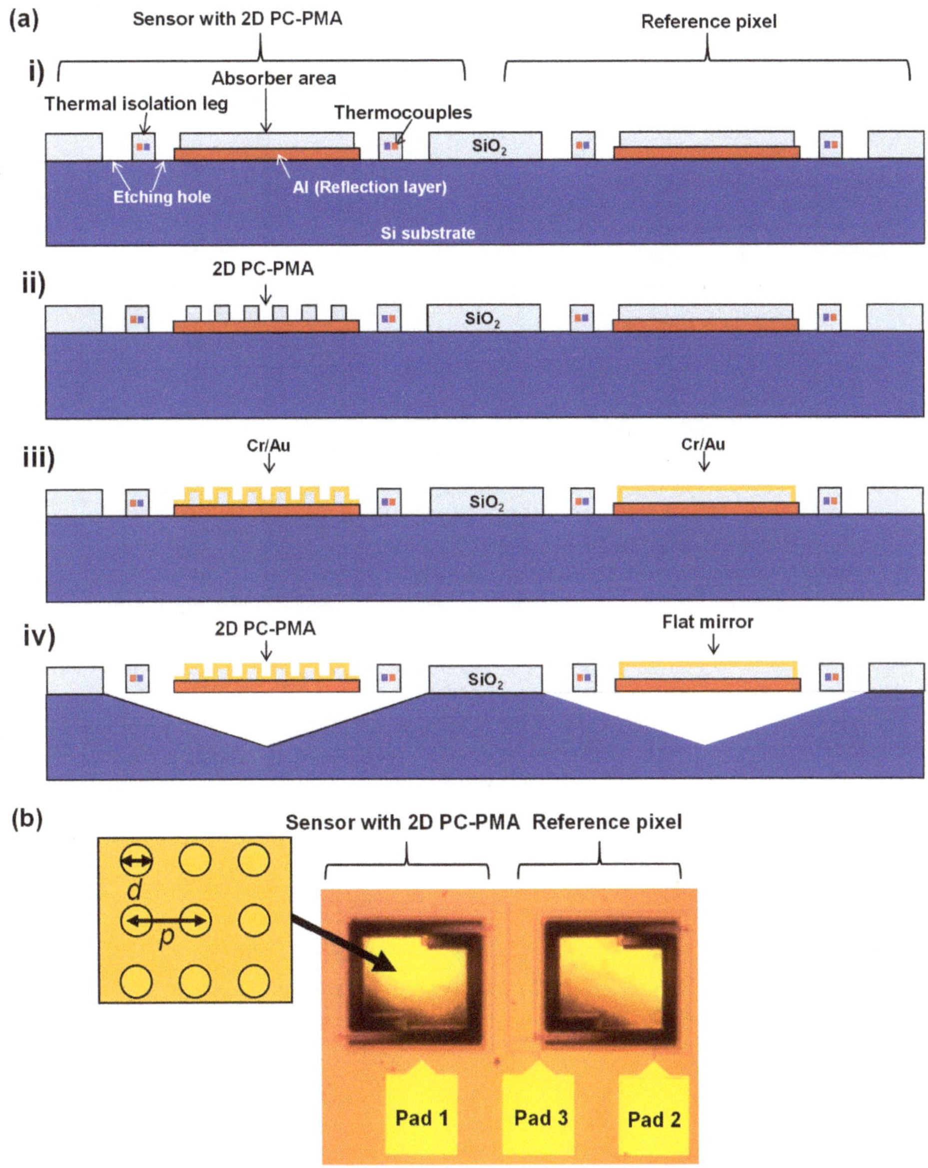

Figure 5. (**a**) Fabrication procedure of a MEMS-based thermopile with PMAs and a reference pixel. (**b**) Optical microscope image of the developed sensor with PMAs and the reference pixel.

Figure 5b shows the optical image of the fabricated sensor with 2D PC-PMA and reference pixel. The absorber area is 300 × 200 μm^2. Pad 3 (in Figure 5b) was added in the actual system for measuring the output voltage of the sensor with the PMAs (V_S) and the reference pixel (V_R) independently, as shown in Figure 3b. It should be noted that the surface pattern of the 2D PC-PMA is too fine to be observed using an optical microscope. The diameter and the period of the dimples in 2D PC-PMA were defined as d and p, respectively. The depth of the dimples was fixed at 1.5 μm in this study.

We fabricated three kinds of the 2D PC-PMA sensors with d and p values of 4.0 and 5.0 μm, 4.0 and 5.5 μm, 4.0 and 6.0 μm, 4.0 and 6.5 μm, and 7.0 and 8.0 μm, which are labeled as sensors A, B, C, D, and E, respectively. The same reference pixel was used for comparison with each sensor. The sensors, A, B, C, D, and E, were expected to have wavelength-selective absorptions at 5.0, 5.5, 6.0, 6.5, and 8.0 μm, respectively, depending on their surface period.

4. Measurement and Results

To measure the spectral responsivity, the sensors with 2D PC-PMAs and the reference pixel were set in a vacuum chamber with a Ge window. The sensors with 2D PC-PMAs and the reference pixel were irradiated simultaneously with IR radiation from a blackbody at a temperature of 1000 K through a narrow bandpass filter. The incidence angle was normal to the 2D PC-PMA sensor and the reference pixel. The output voltages such as V_S, V_R, and ($V_S - V_R$) were monitored using a computer. The responsivity (V/W) was calculated as the ratio between the difference in the output voltage for the on and off states and the input power. The measurement system used in this study is the same as the one reported in our previous studies [6–9,20], except for the use of pinholes included in this study. The input power was calculated from the spectral radiant emittance equation at the evaluated wavelength, with the measurement system parameters such as transmittance of the IR ray power from the blackbody to the sensor through the air, narrow bandpass filters and the Ge window, and the absorber area of the sensors, as previously reported [6–9,20].

Figure 6a,b shows the normalized spectral responsivity of sensor B and the reference pixel, and the normalized spectral responsivity after the subtraction operation, respectively. It is worth noting that the responsivity was normalized with respect to the maximum value in order to avoid any misleading results because the absorption peak caused by surface plasmon resonance was so sharp that wavelength resolution of the measurement system was not sufficient to precisely determine the maximum value. Nevertheless, we confirmed that the developed sensor exhibited comparable responsivity to conventional thermopiles. For example, a maximum responsivity of 100 V/W was achieved. Figure 6a shows clearly that the unwanted modes were produced in the LWIR region, and the wavelength-selective detection was degenerated by these unwanted modes. The wavelength-selective detection was clearly observed at 5.5 μm after the subtraction operation, as evident in Figure 6b. Figure 6c–e show the normalized spectral responsivity of sensors A, C, D, and E with the subtraction operation, and the wavelength-selective detection was clearly observed at approximately 5.0 μm, 6.0 μm, 6.5 μm, and 8.0 μm, respectively. It should be noted that the normalized responsivity peak of sensor C can be considered to be at 6.0 μm because the peak wavelength can be determined at between 5.5 μm and 6.5 μm due to the asymmetric shape of the peak. The five detected peak wavelengths for each sensor are approximately equal to the *p*-values, which corroborates the theoretical results [6,7]. These results present direct evidence that the unwanted modes were successfully eliminated using the subtraction operation with the reference pixel.

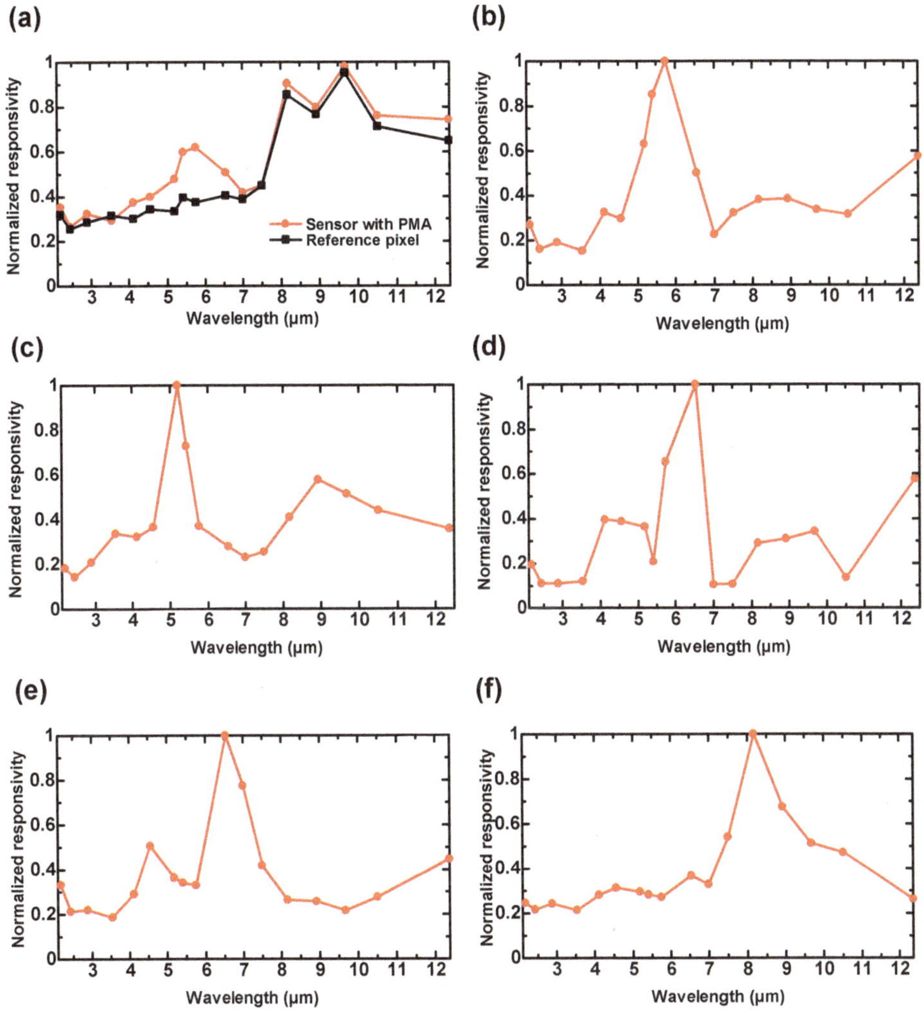

Figure 6. Normalized spectral responsivity: (**a**) of the sensor with PMA and the reference pixel, (**b**) after the subtraction operation for sensor B (p = 5.5 μm and d = 4.0 μm), (**c**) after the subtraction operation for sensor A (p = 5.0 μm and d = 4.0 μm), (**d**) after the subtraction operation for sensor C (p = 6.0 μm and d = 4.0 μm), (**e**) after the subtraction operation for sensor D (p = 6.5 μm and d = 4.0 μm), and (**f**) after the subtraction operation for sensor E (p = 8.0 μm and d = 7.0 μm).

5. Conclusions

In this study, we present a method for the elimination of unwanted modes in a wavelength-selective uncooled IR sensor, using a subtraction operation with a reference pixel. To this end, MEMS-based wavelength-selective uncooled IR sensors using 2D PC-PMAs with reference pixels were fabricated using CMOS and micromachining techniques. The reference pixel possessed the same sensor structure, except a flat mirror was used on the absorber surface instead of PMAs. The spectral responsivity measurements demonstrated that single-mode detection was achieved through the subtraction operation with a reference pixel. It is important to consider the absorptions in other regimes by the PC-PMA-based uncooled IR sensors from an application standpoint. The concept demonstrated in this study can be applied to any type of uncooled IR sensor, such as bolometers [35], silicon-on-insulator diodes [36,37], ferroelectrics [38], and photomechanical sensors [39,40] with wavelength- or polarization-selective absorbers that are capable of multispectral or polarimetric detection, and also to any type of PMAs such as PC, MIM, mushroom, and other types of structures.

Author Contributions: S.O. conceived and designed the experiments and theoretical investigations; Y.T. fabricated the sensors and performed measurements; M.K. supervised all aspects of the research; all authors discussed the results and contributed to writing the manuscript.

Funding: This research received no external funding.

Conflicts of Interest: The authors declare no conflict of interests.

References

1. Vollmer, M.; Mollmann, K.-P. *Infrared Thermal Imaging: Fundamentals, Research and Applications*; Wiley-VCH press: Weinheim, Germany, 2010.
2. Kimata, M. IR imaging. In *Comprehensive Microsystems*; Gianchandani, Y., Tabata, O., Zappe, H., Eds.; Elsevier: Amsterdam, Netherlands, 2008; Volume 3, pp. 113–163.
3. Ogawa, S.; Kimata, M. Wavelength- or polarization-selective thermal infrared detectors for multi-color or polarimetric imaging using plasmonics and metamaterials. *Materials* **2017**, *10*, 493. [CrossRef] [PubMed]
4. Schott, J.R. *Fundamentals of Polarimetric Remote Sensing*; SPIE press: Bellingham, WA, USA, 2009.
5. Gurton, K.P.; Yuffa, A.J.; Videen, G.W. Enhanced facial recognition for thermal imagery using polarimetric imaging. *Opt. Lett.* **2014**, *39*, 3857–3859. [CrossRef] [PubMed]
6. Ogawa, S.; Okada, K.; Fukushima, N.; Kimata, M. Wavelength selective uncooled infrared sensor by plasmonics. *Appl. Phys. Lett.* **2012**, *100*, 021111. [CrossRef]
7. Ogawa, S.; Komoda, J.; Masuda, K.; Kimata, M. Wavelength selective wideband uncooled infrared sensor using a two-dimensional plasmonic absorber. *Opt. Eng.* **2013**, *52*, 127104. [CrossRef]
8. Takagawa, Y.; Ogawa, S.; Kimata, M. Detection wavelength control of uncooled infrared sensors using two-dimensional lattice plasmonic absorbers. *Sensors* **2015**, *15*, 13660–13669. [CrossRef]
9. Ogawa, S.; Takagawa, Y.; Kimata, M. Broadband polarization-selective uncooled infrared sensors using tapered plasmonic micrograting absorbers. *Sens. Actuators A Phys.* **2018**, *269*, 563–568. [CrossRef]
10. Ogawa, S.; Kimata, M. Metal-insulator-metal-based plasmonic metamaterial absorbers at visible and infrared wavelengths: A review. *Materials* **2018**, *11*, 458. [CrossRef]
11. Ogawa, S.; Fujisawa, D.; Hata, H.; Uetsuki, M.; Misaki, K.; Kimata, M. Mushroom plasmonic metamaterial infrared absorbers. *Appl. Phys. Lett.* **2015**, *106*, 041105. [CrossRef]
12. Ogawa, S.; Fujisawa, D.; Hata, H.; Kimata, M. Absorption properties of simply fabricated all-metal mushroom plasmonic metamaterials incorporating tube-shaped posts for multi-color uncooled infrared image sensor applications. *Photonics* **2016**, *3*, 9. [CrossRef]
13. Ogawa, S.; Fujisawa, D.; Kimata, M. Theoretical investigation of all-metal-based mushroom plasmonic metamaterial absorbers at infrared wavelengths. *Opt. Eng.* **2015**, *54*, 127014. [CrossRef]
14. Maier, T.; Brueckl, H. Multispectral microbolometers for the midinfrared. *Opt. Lett.* **2010**, *35*, 3766–3768. [CrossRef]
15. Hao, J.; Wang, J.; Liu, X.; Padilla, W.J.; Zhou, L.; Qiu, M. High performance optical absorber based on a plasmonic metamaterial. *Appl. Phys. Lett.* **2010**, *96*, 251104. [CrossRef]
16. Liu, X.; Starr, T.; Starr, A.F.; Padilla, W.J. Infrared spatial and frequency selective metamaterial with near-unity absorbance. *Phys. Rev. Lett.* **2010**, *104*, 207403. [CrossRef]
17. Matsuno, Y.; Sakurai, A. Perfect infrared absorber and emitter based on a large-area metasurface. *Opt. Mater. Express* **2017**, *7*, 618–626. [CrossRef]
18. Zhang, B.; Zhao, Y.; Hao, Q.; Kiraly, B.; Khoo, I.C.; Chen, S.; Huang, T.J. Polarization-independent dual-band infrared perfect absorber based on a metal-dielectric-metal elliptical nanodisk array. *Opt. Express* **2011**, *19*, 15221–15228. [CrossRef]
19. Cheng, C.W.; Abbas, M.N.; Chiu, C.W.; Lai, K.T.; Shih, M.H.; Chang, Y.C. Wide-angle polarization independent infrared broadband absorbers based on metallic multi-sized disk arrays. *Opt. Express* **2012**, *20*, 10376–10381. [CrossRef]

20. Ogawa, S.; Takagawa, Y.; Kimata, M. Fano resonance in asymmetric-period two-dimensional plasmonic absorbers for dual-band uncooled infrared sensors. *Opt. Eng.* **2016**, *55*, 117105. [CrossRef]
21. Zhao, L.; Niu, Q.; He, Z.; Yang, X.; Dong, S. Theoretical excitation of 2-D (1, 1) cavity mode with asymmetric sword-shaped notched square resonators for metamaterial perfect multiband absorbers in infrared range. *Opt. Express* **2018**, *26*, 31510–31522. [CrossRef]
22. Zhao, L.; Liu, H.; He, Z.; Dong, S. Theoretical design of twelve-band infrared metamaterial perfect absorber by combining the dipole, quadrupole, and octopole plasmon resonance modes of four different ring-strip resonators. *Opt. Express* **2018**, *26*, 12838–12851. [CrossRef]
23. Aslan, E.; Aslan, E.; Turkmen, M.; Saracoglu, O.G. Metamaterial plasmonic absorber for reducing the spectral shift between near- and far-field responses in surface-enhanced spectroscopy applications. *Sens. Actuators A Phys.* **2017**, *267*, 60–69. [CrossRef]
24. Aydin, K.; Ferry, V.E.; Briggs, R.M.; Atwater, H.A. Broadband polarization-independent resonant light absorption using ultrathin plasmonic super absorbers. *Nat. Commun.* **2011**, *2*, 517. [CrossRef]
25. Bouchon, P.; Koechlin, C.; Pardo, F.; Haïdar, R.; Pelouard, J.L. Wideband omnidirectional infrared absorber with a patchwork of plasmonic nanoantennas. *Opt. Lett.* **2012**, *37*, 1038–1040. [CrossRef]
26. Liu, T.; Takahara, J. Ultrabroadband absorber based on single-sized embedded metal-dielectric-metal structures and application of radiative cooling. *Opt. Express* **2017**, *25*, A612–A627. [CrossRef]
27. Lei, L.; Li, S.; Huang, H.; Tao, K.; Xu, P. Ultra-broadband absorber from visible to near-infrared using plasmonic metamaterial. *Opt. Express* **2018**, *26*, 5686–5693. [CrossRef]
28. Shrestha, S.; Wang, Y.; Overvig, A.C.; Lu, M.; Stein, A.; Negro, L.D.; Yu, N. Indium tin oxide broadband metasurface absorber. *ACS Photonics* **2018**, *5*, 3526–3533. [CrossRef]
29. Sabah, C.; Esat, V.; Obaidullah, M. Polarization independent triple-band (5,4) semiconducting carbon nanotube metamaterial absorber design for visible and ultraviolet regions. *J. Nanophoton.* **2017**, *11*, 046011.
30. Yahiaoui, R.; Hanai, K.; Takano, K.; Nishida, T.; Miyamaru, F.; Nakajima, M.; Hangyo, M. Trapping waves with terahertz metamaterial absorber based on isotropic Mie resonators. *Opt. Lett.* **2015**, *40*, 3197–3200. [CrossRef]
31. Tao, H.; Landy, N.I.; Bingham, C.M.; Zhang, X.; Averitt, R.D.; Padilla, W.J. A metamaterial absorber for the terahertz regime: Design, fabrication and characterization. *Opt. Express* **2008**, *16*, 7181–7188. [CrossRef]
32. Ogawa, S.; Takagawa, Y.; Kimata, M. Elimination of unwanted modes in wavelength-selective uncooled infrared sensors using plasmonic metamaterial absorbers. In *Infrared Technology and Applications XLV, Proceedings of SPIE Defense + Commercial Sensing. Baltimore, MD, United States, 17 April 2019*; SPIE: Bellingham, WA, USA, 2019; Volume 11002, p. 110021K.
33. Hirota, M.; Satou, F.; Saito, M.; Kishi, Y.; Nakajima, Y.; Uchiyama, M. Thermoelectric infrared imager and automotive applications. In *Infrared Technology and Applications XXVII, Proceedings of Aerospace/Defense Sensing, Simulation, and Controls, Orlando, FL, USA, October 2001*; SPIE: Bellingham, WA, USA, 2001; Volume 4369, pp. 312–321.
34. Moharam, M.G.; Gaylord, T.K. Rigorous coupled-wave analysis of planar-grating diffraction. *J. Opt. Soc. Am.* **1981**, *71*, 811–818. [CrossRef]
35. Wood, R.A. High-performance infrared thermal imaging with monolithic silicon focal planes operating at room temperature. In *IEDM, Proceedings of IEEE International Electron Devices Meeting, Washington, DC, USA, 5–8 December 1993*; IEEE: Piscatway, NJ, US, 1993; pp. 175–177.
36. Ishikawa, T.; Ueno, M.; Nakaki, Y.; Endo, K.; Ohta, Y.; Nakanishi, J.; Kosasayama, Y.; Yagi, H.; Sone, T.; Kimata, M. Performance of 320 x 240 uncooled IRFPA with SOI diode detectors. In *Infrared Technology and Applications XXVI, Proceedings of International Symposium on Optical Science and Technology, San Diego, CA, USA, December 2000*; SPIE: Bellingham, WA, USA, 2000; Volume 4130, pp. 152–159.
37. Maegawa, T.; Fujisawa, D.; Hata, H.; Ogawa, S.; Ueno, M. 2-in-1 diodes with a contact-sidewall structure for small pixel pitch in silicon-on-insulator (SOI) uncooled infrared (IR) focal plane arrays. *Sens. Mater.* **2014**, *26*, 189–198.
38. Belcher, J.F.; Hanson, C.M.; Beratan, H.R.; Udayakumar, K.R.; Soch, K.L. Uncooled monolithic ferroelectric IRFPA technology. In *Infrared Technology and Applications XXIV, Proceedings of SPIE's International Symposium on Optical Science, Engineering, and Instrumentation, San Diego, CA, USA, October 1998*; SPIE: Bellingham, WA, USA, 1998; Volume 3436, pp. 611–622.

39. Zhao, J. High sensitivity photomechanical MW-LWIR imaging using an uncooled MEMS microcantilever array and optical readout. In *Infrared Technology and Applications XXXI, Proceedings of Defense and Security, Orlando, FL, USA, May 2005*; SPIE: Bellingham, WA, USA, 2005; Volume 5783, pp. 506–513.
40. Hui, Y.; Gomez-Diaz, J.S.; Qian, Z.; Alù, A.; Rinaldi, M. Plasmonic piezoelectric nanomechanical resonator for spectrally selective infrared sensing. *Nat. Commun.* **2016**, *7*, 11249. [CrossRef]

materials

MDPI

Review

Metal-Insulator-Metal-Based Plasmonic Metamaterial Absorbers at Visible and Infrared Wavelengths: A Review

Shinpei Ogawa [1,*] and Masafumi Kimata [2]

1 Advanced Technology R&D Center, Mitsubishi Electric Corporation, 8-1-1 Tsukaguchi-Honmachi, Amagasaki, Hyogo 661-8661, Japan

2 College of Science and Engineering, Ritsumeikan University, 1-1-1 Noji-higashi, Kusatsu, Shiga 525-8577, Japan; kimata@se.ritsumei.ac.jp

* Correspondence: Ogawa.Shimpei@eb.MitsubishiElectric.co.jp; Tel.: +81-6-6497-7533

Received: 23 February 2018; Accepted: 17 March 2018; Published: 20 March 2018

Abstract: Electromagnetic wave absorbers have been investigated for many years with the aim of achieving high absorbance and tunability of both the absorption wavelength and the operation mode by geometrical control, small and thin absorber volume, and simple fabrication. There is particular interest in metal-insulator-metal-based plasmonic metamaterial absorbers (MIM-PMAs) due to their complete fulfillment of these demands. MIM-PMAs consist of top periodic micropatches, a middle dielectric layer, and a bottom reflector layer to generate strong localized surface plasmon resonance at absorption wavelengths. In particular, in the visible and infrared (IR) wavelength regions, a wide range of applications is expected, such as solar cells, refractive index sensors, optical camouflage, cloaking, optical switches, color pixels, thermal IR sensors, IR microscopy and gas sensing. The promising properties of MIM-PMAs are attributed to the simple plasmonic resonance localized at the top micropatch resonators formed by the MIMs. Here, various types of MIM-PMAs are reviewed in terms of their historical background, basic physics, operation mode design, and future challenges to clarify their underlying basic design principles and introduce various applications. The principles presented in this review paper can be applied to other wavelength regions such as the ultraviolet, terahertz, and microwave regions.

Keywords: plasmonics; metamaterials; metal-insulator-metal; absorbers

---

1. Introduction

Electromagnetic (EM) wave absorbers are drawing significant interest from aspects of both fundamental science and industry applications. Typical EM wave absorbers are essentially based on the intrinsic loss of the material and thus require a long optical path, which results in large volume and poor design flexibility. EM wave absorbers with absorption properties that can be efficiently controlled by their structures have thus been studied for many years. Such EM wave absorbers were first studied in the microwave range and are roughly classified into two groups, according to Reference [1], as broadband absorbers and resonant absorbers. The broadband absorbers are further categorized into two groups: geometric transition absorbers and low-density absorbers [1]. Geometric transition absorbers consist of two-dimensional (2D) periodic pyramids that cause a gradual change in the dielectric constant from the free space to the absorbers [2,3]. Low-density absorbers utilize porous materials [4,5] and the multi-reflections that occur in these pores has led to significant absorption, which was realized using thin absorbers.

The resonance absorbers are classified into three types, according to Reference [6]. Figure 1a–f shows schematic illustrations and the reflectance of the Salisbury screen, Jaumann absorber, and circuit analog (CA) absorber. All of these resonance absorbers use a quarter-wavelength gap from the top material to the bottom substrate. The Salisbury screen uses a non-periodic resistive sheet in front of a ground plate [7]. The Jaumann absorber uses two or more resistive sheets in front of each other and is a basic resonance absorber [8]. These two absorbers use purely resistive sheets. The CA absorber uses a periodic surface made of a lossy material with three layers: the top periodic metal patterns, a middle dielectric layer, and a continuous metallic bottom layer [6]. The concept of CA absorbers is the basis of recent metamaterial absorbers for a wide range of wavelength regions, from visible to microwave wavelengths.

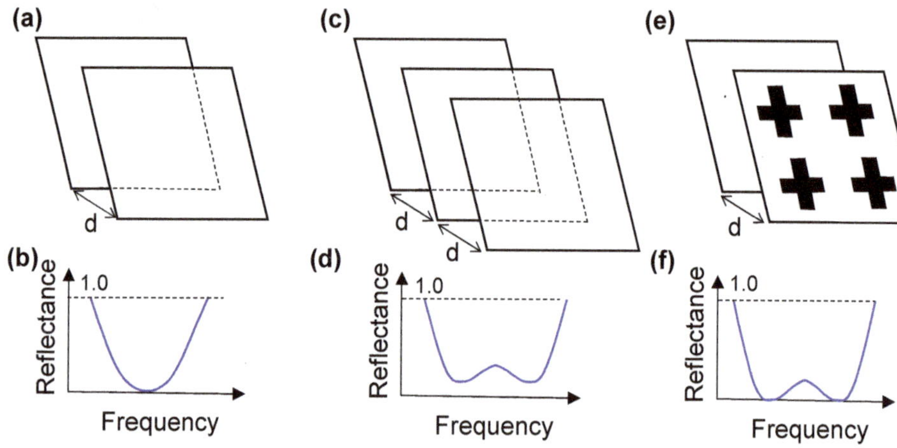

Figure 1. Schematic illustrations and reflectance of resonant absorbers: (**a**,**b**) the Salisbury screen; (**c**,**d**) the Jaumann absorber; and (**e**,**f**) the circuit analog CA absorber. "d" represents the quarter-wavelength gap [6].

Recent advances in plasmonics [9] and metamaterials [10] research together with the progress in nanotechnological fabrication techniques has led to novel EM absorbers at visible and infrared (IR) wavelengths. These absorbers uses localized surface plasmon polaritons (LSPPs) [11] with a metamaterial concept to achieve much smaller absorber volumes, sufficient performance, and design flexibility based on geometry rather than the materials used. SPPs are the collective oscillation of electrons between metals and dielectrics that can go beyond the diffraction limit [12]. LSPPs are key to realizing small and thin absorbers for the visible and IR wavelength regions. Therefore, much significant research has been performed on EM wave absorbers using SPPs or LSPPs at visible and IR wavelengths.

There are roughly two categories of absorbers that employ plasmonics and metamaterials: conventional periodic structures such as plasmonic crystals [13–15] and gratings [16–19], and metamaterial-based structures, where periodicity has less impact on the optical properties [20]. In particular, metal-insulator-metal-based plasmonic metamaterial absorbers (MIM-PMAs) are the most promising and widely studied for a wide range of wavelengths due to their high performance, such as high absorbance, incident angle, and polarization insensitivity, as well as their design flexibility and simple fabrication. Although their fundamental principles are basically the same, a wide range of applications is expected, such as solar cells [21], refractive index sensors [22], optical camouflage [23], cloaking [24], optical switches [25], color pixels [26,27], thermal IR sensors [28–31], mechanical thermal sensors [32], surface-enhanced spectroscopy [33,34], and gas sensing [35]. Therefore, this review paper aims to clarify the fundamental principle, characteristics, possibilities, and challenges of MIM-PMAs at visible and IR wavelengths to contribute to future research and the expansion of their applications.

Please note that MIM-based thermal emitters are considered as MIM-PMAs at IR wavelengths [36] because absorbance is equal to emissivity, as given by Kirchhoff's law. To the best of our knowledge, MIM-PMAs were first demonstrated as thermal IR emitters by Puscasu and Schaich [37].

2. Structures and Materials

In this section, the fundamental structures and materials of MIM-PMAs are explained with a simple introduction of the basic optical properties. The detailed optical characteristics are discussed in a later section.

MIM-PMAs consist of three layers: a bottom metal layer, a middle dielectric layer, and a top periodic metal micropatches. Figure 2a,b shows schematic illustrations of conventional MIM-PMAs with two-dimensional (2D) and one-dimensional (1D) periodic micropatches, respectively. The absorption wavelength is fundamentally defined by the micropatch size. The 2D configuration is symmetric for two orthogonal directions, the x and y directions; therefore, the optical properties are polarization insensitive. On the other hand, the 1D configuration is asymmetric in the x and y directions, so the optical properties are polarization sensitive. Figure 2c,d shows cross-sectional views of the conventional MIM-PMAs with flat and isolated dielectric layers, respectively. The middle dielectric layers underneath the micropatches are required, so that both structures function as MIM-PMAs.

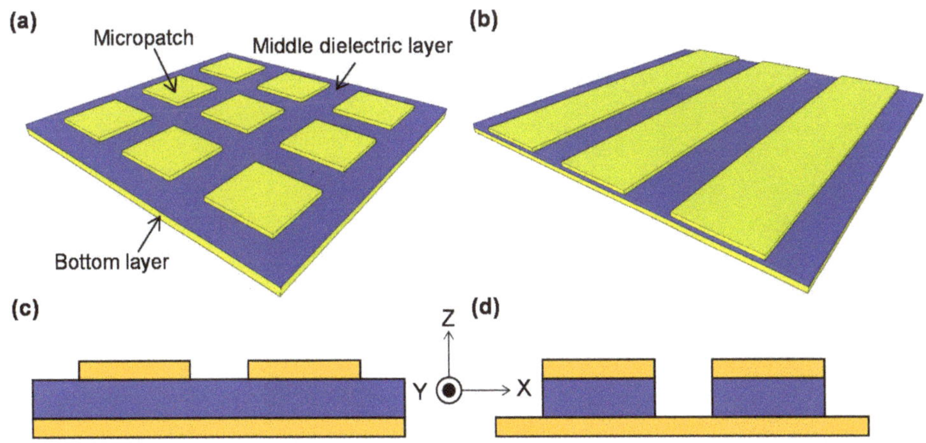

Figure 2. Schematic illustrations of metal-insulator-metal-based plasmonic metamaterial absorbers (MIM-PMAs); oblique view of (**a**) two-dimensional (2D) and (**b**) one-dimensional (1D) periodic micropatches. Cross-sectional views of (**c**) continuous and (**d**) isolated middle dielectric layers.

The thickness of the metal in the bottom layer and the top micropatches is required to be more than twice the depth of the operating wavelength region, e.g., 100 nm thickness is sufficient for the IR wavelength region [38]. The thickness of the middle dielectric layer can be less than the operating wavelength/50 due to the strong confinement of the waveguide mode of SPPs [39].

The possible lattice structures for 2D periodic micropatches are square, triangular, and honeycomb. However, the lattice structures have less impact on absorption properties because each micropatch acts as a single isolated resonator [40].

The shape of micropatches are roughly classified into symmetric in the two orthogonal directions, such as squares [28,41], circles [42], and crosses [38,43], and nano-particles [44–46], or asymmetric, such as ellipses [47], rectangles [48], wedges [49], bow-ties [50], split-ring resonators [51], and asymmetric crosses [52,53]. Nanocubes have also been used as micropatches [54,55]. The shape of the corners and the sidewall angles have an important role in defining single and multiband resonances [56]. The end shape of the micropatch produces the difference of optical modes formed in the middle dielectric layer because MIM structures can be considered as waveguides and the shape of the waveguide end defines the waveguide mode [57]. The first three micropatch shapes are symmetric in the x and y directions, and absorption occurs at a single wavelength. On the other hand, the latter four micropatch shapes are asymmetric in the x and y directions and produce two absorption wavelengths. As discussed in Section 4, the symmetry is an important parameter for polarization dependence.

It is also important to consider the temperature tolerance and compatibility of the complementary metal oxide semiconductor (CMOS) process for the choice of materials used for MIM-PMAs [43,58]. Tables 1 and 2 show the properties of the metals and dielectrics used in MIM-PMAs [43].

The top and the bottom layers are typically based on metals such as gold (Au) [38,41], silver (Ag) [49], and aluminum (Al) [28], which are common materials for SPPs. Titanium nitride (TiN) [59,60], molybdenum (Mo) [43], and tungsten (W) [58], or highly-doped silicon (Si) [61], can be used for the bottom and top micropatches. Graphene can also be used for top micropatches in the IR wavelength region [62]. TiN or Mo have recently been used for thermal IR emitters due to requirements of high-temperature tolerance. However, absorbers require less temperature tolerance. Therefore, Al is widely employed due to its compatibility with the CMOS process and its low cost.

Table 1. Properties of metals used in MIM-PMAs. Adapted with permission from Reference [43]. © 2017 American Chemical Society. (CMOS: complementary metal oxide semiconductor).

Material	Melting Point (°C)	Electrical Conductivity ($\times10^7$ S/m @20 °C)	Plasma Frequency ($\times10^{15}$ Hz)	Thermal Expansion Coefficient (CTE; $\times10^{-6}$ K^{-1})	Young's Modulus (E; GPa)	CMOS Compatibility
Al	660	3.5	3.57	24	70	yes
Au	1000	4.52	2.2	14	78	no
Pt	1770	0.944	1.25	8.8	168	no
TiN	2930	0.87	1.84	9.35	251	yes
Mo	2620	1.9	1.8	4.8	329	yes

Table 2. Properties of insulators used in MIM-PMAs. Adapted with permission from Reference [43]. © 2017 American Chemical Society.

Material	Melting Point (°C)	Young's Modulus (E; GPa)	Poisson Ratio (μ)	CTE ($\times10^{-6}$ K^{-1})	Thermal Conductivity (W/m·K)
AlN	2200	344.8	0.287	4.6	285
SiO_2	1600	70	0.17	0.5	1.4
Al_2O_3	2072	353.1	0.22	4.5	25.08

The middle layer is roughly classified into two groups: oxides or nitrides such as Al_2O_3 [41], SiO_2 [40], CeO_2 [63], SiN [28], and AlN [43], and semiconductors such as Si [64], germanium (Ge) [65], and ZnS [66]. Another material often used is MgF_2 [33]. Phase transition materials of germanium antimony telluride [67] and VO_2 [68] have been used for thermal switching. The loss is an important factor of the middle dielectric layer, as discussed in a later section. Lossless materials such as Si, Ge, ZnS, and CeO_2 should be used for the IR wavelength region in the vicinity of 10 μm to maintain the linear tunability of the absorption wavelength. However, these materials are not compatible with the CMOS process. The alternative choice to avoid the influence of intrinsic loss materials is mushroom-PMAs, where small post structures connect the micropatches and the bottom layer without a continuous middle dielectric layer [69–71].

A perforated metal plate can be used as a top layer, which is a complementary structure of periodic micropatches [72] that provides optical properties similar to MIM-PMAs. However, there is less design flexibility due to the need for periodicity. 1D grating structures with ultra-narrow groove widths (ca. 100 nm) and high aspect ratios (>10) [16,18,73] can be classified as MIM-PMAs because such a slit is considered to be the waveguide, which is equivalent to the insulator layer in MIM structures [74,75]. However, high aspect ratios and narrow groove widths require complicated fabrication procedures. In this paper, these structures are excluded and instead focus is made on the conventional MIM-PMAs, shown as Figure 2. Other structures such as core-shell nanoparticles [76] and multi-flat-layer structures [77–79] can be considered to have a principle similar to that of MIM-PMAs.

3. Basic Optical Properties

3.1. Principle

Figure 3a shows the operating principle of MIM-PMAs for incident EM waves. Figure 3b–d shows the calculated electric and magnetic fields, and the power distribution of MIM-PMAs at the absorption wavelength, respectively [41].

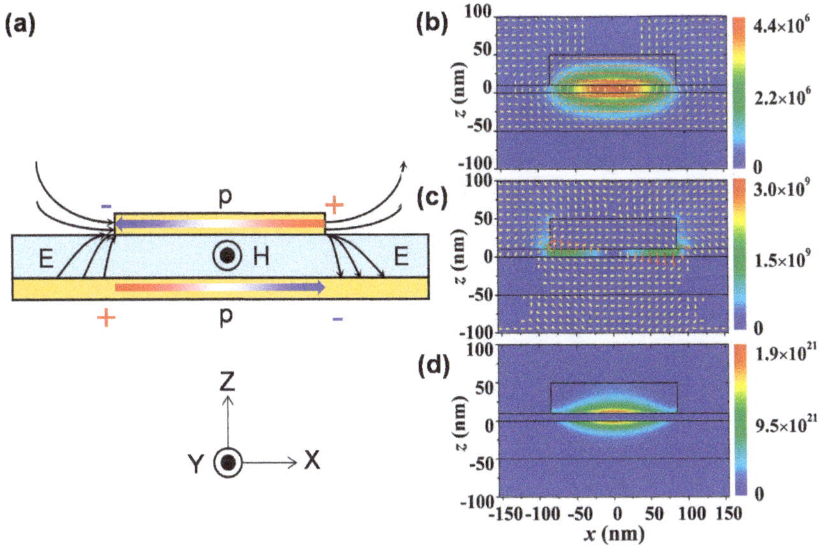

Figure 3. (**a**) Schematic illustration of the operating principle of MIM-PMAs. E, H, and p represent the electric displacement vector, magnetic field, and current, respectively. Calculated results: (**b**) magnetic field; (**c**) electric field; and (**d**) power distribution. The color maps represents the amplitude of each distribution. (**b–d**) are reprinted from Reference [41] with the permission of AIP Publishing.

As shown in Figure 3a, a pair of anti-parallel oscillating currents is induced in both the bottom layer and the top micropatches, and significant magnetic resonance is produced. Dipole electric resonance is formed accordingly between the edge of the micropatches and the near bottom layer. LSPPs are induced by the incident light at the absorption wavelength. This principle is confirmed by the calculated electromagnetic field distribution, as shown in Figure 3b–d [41]. The electric displacement vectors in the bottom layer and the micropatches are opposite to each other, which generates a strong magnetic response [80] (Figure 3b). Strong electric dipole resonances are observed at the sides of the micropatches (Figure 3c). The reflectance can be completely cancelled in the far field by the interference of these two dipoles due to the π shift phase [65]. Strong absorption is thus attributed to these localized magnetic and electric dipole resonances, which provide sufficient time to consume light energy by the ohmic losses in the metals (Figure 3d).

3.2. Wavelength Selectivity

Figure 4a,b shows the calculated and measured optical properties of MIM-PMAs in IR wavelengths [38]. Wavelength selective absorption is clearly obtained at 6 μm, which is a typical wavelength-selective absorption property of MIM-PMAs. The main absorption wavelength is always longer than the period of the micropatches because wavelengths smaller than the period are diffracted. Figures 4c [41] and 4d [81] show the calculated absorbance as a function of the wavelength and the micropatch size (w) in the near-IR wavelength region, as well as the measured relation between the micropatch size and the absorption wavelength in the IR wavelength region.

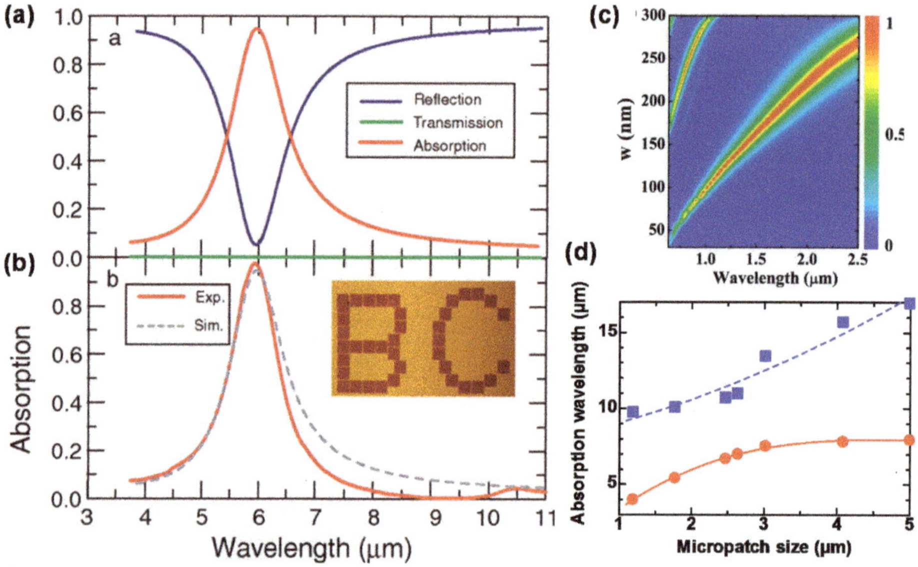

Figure 4. Calculated spectra for (**a**) absorption, reflection, and transmission; (**b**) Comparison of the measured and calculated absorption spectra; (**c**) Absorbance as a function of the wavelength and the micropatch size (w) in the near-IR wavelength region; (**d**) Relation between the micropatch size and the absorption wavelength in the IR wavelength region. (**a**,**b**) are adapted with permission from Reference [38]. © 2010 American Physical Society. (**c**) is reprinted from Reference [41] with the permission of AIP Publishing.

The absorption wavelength is almost proportional to the micropatch size in the near-IR wavelength region. In contrast, the relation between the micropatch size and the absorption wavelength is non-linear in the IR wavelength region, which is attributed to the loss of the middle dielectric layer [40]. Most oxides become lossy in the vicinity of 10 μm, where no absorption occurs, and this causes the non-linearity between the micropatch size and the absorption wavelength, as shown in Figure 4d. This is an important point for the design of MIM-PMAs for use at IR wavelengths. The thickness of the middle dielectric layer has less impact on the absorbance and the absorption wavelength, and can thus be optimized for the operating wavelength [41].

3.3. Incidence Angle Dependence

Figure 5a,b shows the calculated incident angle dependence of the absorbance as a function of the wavelength for the transverse-electric (TE) and transverse-magnetic (TM) modes, respectively [22]. The calculated model is for an MIM-PMA with 2D circle-shaped micropatches.

Figure 5 shows that the absorption can be realized at almost the same wavelength for a wide range of incidence angle up to approximately 70° for both TE and TM modes. This property is attributed to the strong LSPPs, as shown in Figure 3. The incident angle independence is an important advantage for device applications such as solar cells, IR image sensors, and biological sensors.

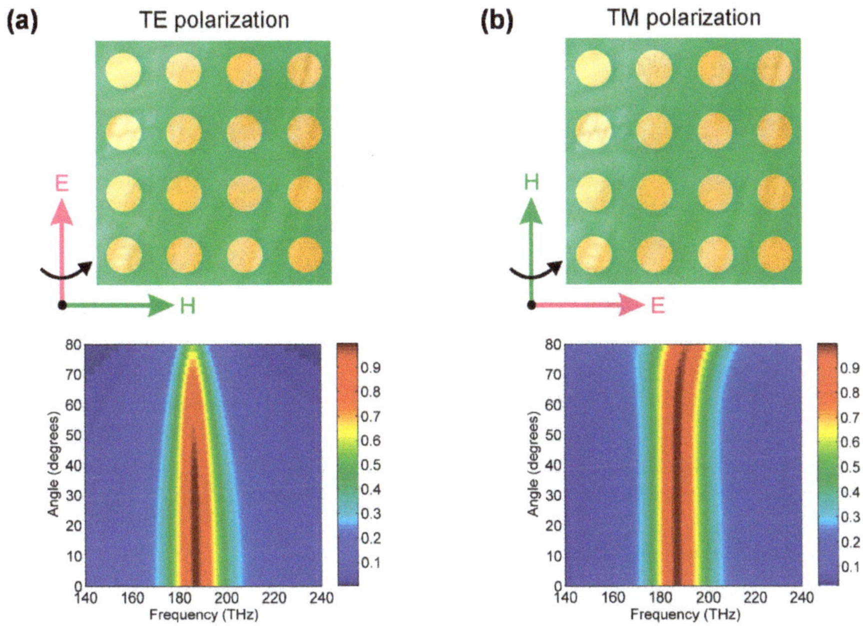

Figure 5. Calculated absorption for the (**a**) transverse-electric (TE) and (**b**) transverse-magnetic (TM) modes. Figures are adapted with permission from Reference [22]. © 2010 American Chemical Society.

3.4. Polarization Dependence

In this section, the coordinate system is set as shown in Figure 3c,d. When the electric field of the incident light is in the x or y direction, the absorption wavelength is defined by the side-length of the micropatches in the x or y direction, respectively [39]. Each side-length of the square-shaped micropatches in the x and y direction is the same. Thus, the absorption wavelength is also the same for each polarization. MIM-PMAs with this configuration are polarization insensitive.

Different side lengths, such as ellipse [47] or asymmetric-cross-shaped [52] micropatches, produce dual band absorption, as discussed in the next section. The 1D periodic configuration shown in Figure 3b also produces two absorption modes. However, when one side-length is much longer than the other, the other absorbance is outside the operation wavelength region, which results in polarization-selective absorbers. Polarization-selective absorbers can be applied to IR polarimetric imaging [82,83] to enhance object recognition ability such as distinct human trace in a natural environment and human facial recognition [84].

3.5. Inductor-capacitor (LC) Circuit Model

The operation principle of MIM-PMAs is sometimes explained using the LC equivalent circuit model. This may be due to the influence of CA absorbers mentioned in the introduction section. Figure 6 shows a schematic illustration of the LC equivalent circuit for MIM-PMAs [40,63]. Two models are considered, with or without the loss of the insulator layer. The frequency that gives a total impedance of zero is the absorption frequency.

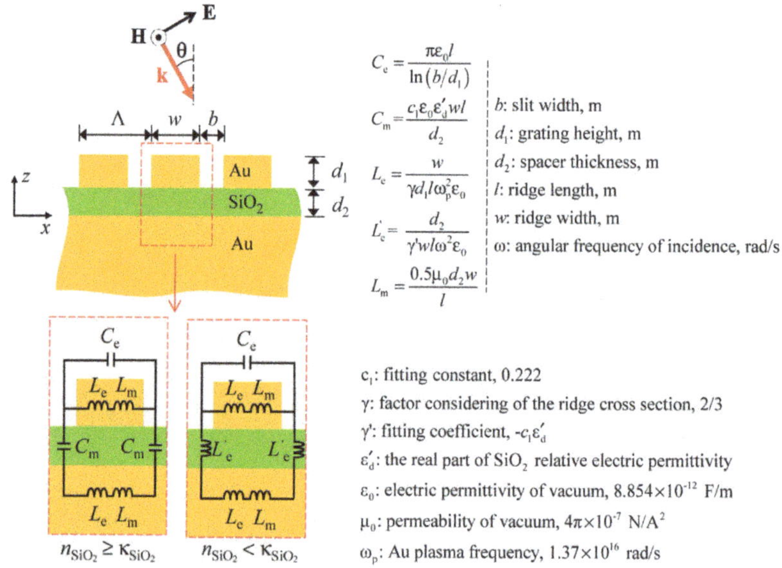

Figure 6. Schematic illustration of the LC equivalent circuit for MIM-PMAs. Figures were adapted with permission from Reference [40]. © 2013 Optical Society of America.

4. Multi-Band and Broadband Operation

The strategies of multi-band and broadband absorption are classified into three groups: asymmetrically-shaped [47,52,85–89] or multi-size [42,65,66,90–94] micropatches, multi-layers of MIM structure [95–100], and embedded in dielectric materials [57,101,102]. The first two are based on multi-resonance that produces multi-mode absorption. Each absorption mode becomes close, which results in broadband absorption [103]. Figure 7a–d shows MIM-PMAs with two-types of asymmetrically shaped micropatches, such as cross [52] and elliptical shapes [47], for dual-band operation. This dual-band absorption is designed in consideration of the polarization dependence for the two orthogonal directions.

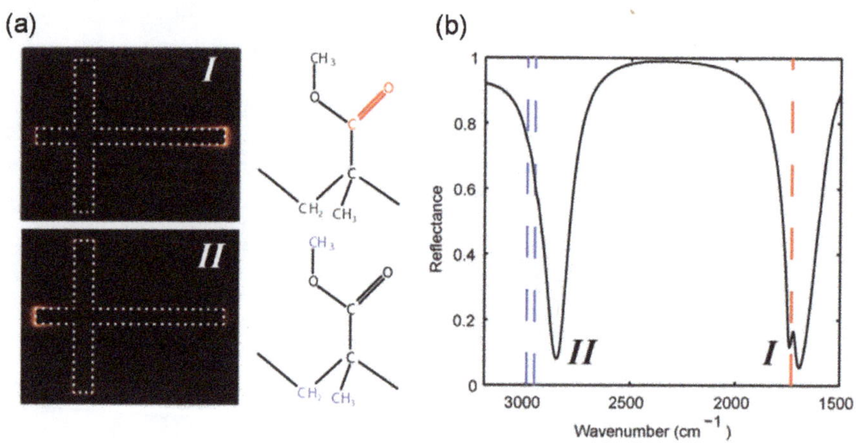

Figure 7. *Cont.*

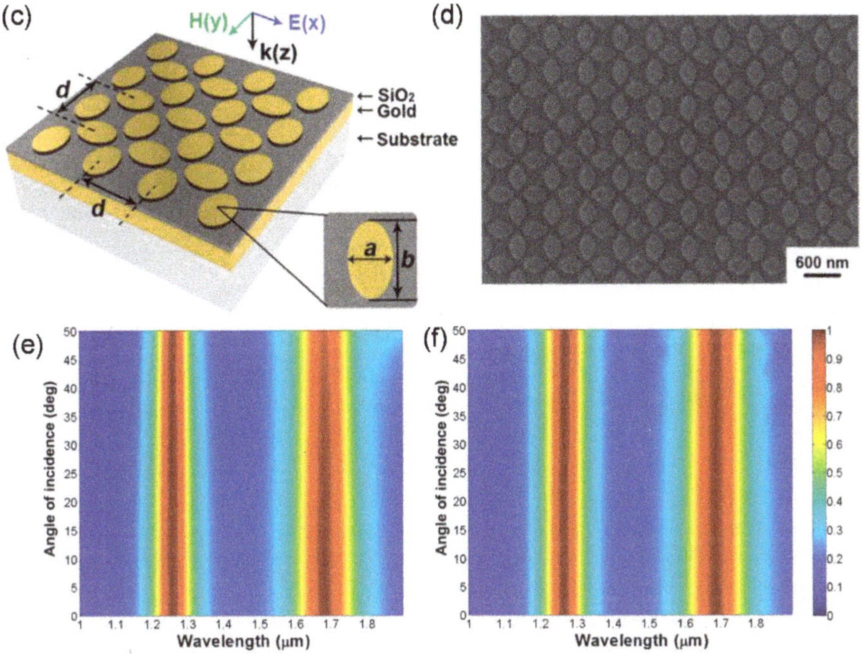

Figure 7. Schematic illustration of an MIM-PMA with asymmetric cross-shaped micropatches for dual-band operation. (**a**) Electric field distribution for modes I and II; and (**b**) the corresponding reflectance spectrum. (**c**) Schematic illustration and (**d**) SEM image of an MIM-PMA with an elliptical nanodisk array. Calculated absorbance for (**e**) TE and (**f**) TM modes. (**a**,**b**) are adapted with permission from Reference [52]. © 2012 American Chemical Society. (**c**–**f**) are adapted with permission from Reference [47]. © 2011 Optical Society of America.

Figure 8a,b shows oblique and the cross-sectional schematic illustrations of an MIM-PMA with 1D stripe-shaped multi-size-micropatches (w_1 to w_4), respectively [65]. Figure 8c,d shows a schematic illustration and magnetic field distribution of an MIM-PMA with 2D multi-size micropatches and the corresponding absorption spectrum, respectively [66]. The absorption spectrum is the summation of the absorption wavelengths generated by each micropatch resonator. Figure 8d shows that the distance between each resonant wavelength becomes close, which results in broadband absorption.

Figure 9a,b shows a schematic illustration of a multi-layer MIM-PMA and the calculated absorption spectrum, respectively [95]. Each MIM layer produces multi-plasmonic-resonances at different wavelengths and each resonance is coupled, so that broadband absorption occurs [103].

Broadband absorption is also achieved by MIM-PMAs with single or multi-layers embedded in lossy dielectrics such as amorphous Si [57] and SiN [101]. The resonances of MIM-PMAs can be broadened by lossy materials, which results in broad absorption. However, these structures increase the thickness and volume of the absorbers, and cause difficulties for fabrication. Care should be taken in applying them to practical devices by comparison with other convenient structures such as simple multi-flat-layer structures [77–79] or gold black [104,105] in terms of their thickness, ease of fabrication, and cost.

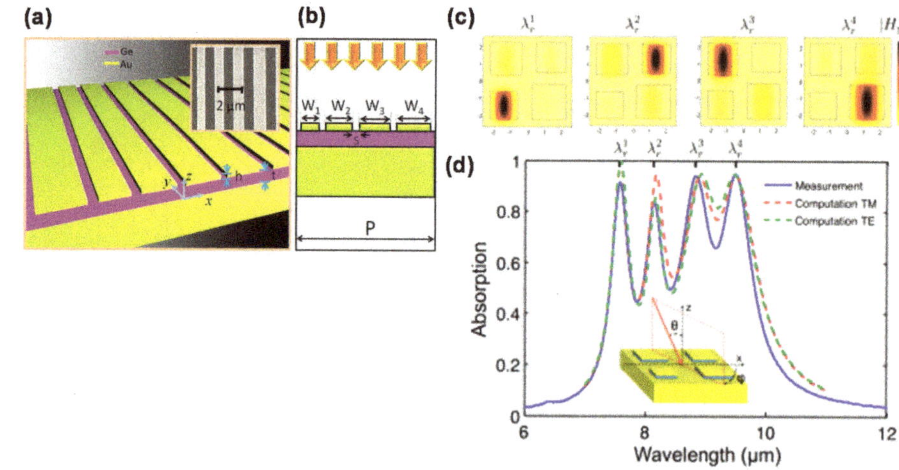

Figure 8. Schematic illustrations of MIM-PMAs with multi-size micropatches having (**a**,**b**) 1D and (**c**) 2D periodic configurations. (**d**) Calculated and measured absorption spectra for the 2D periodic configuration. (**a**,**b**) are reprinted from Reference [65] with the permission of AIP Publishing. (**c**,**d**) are adapted with permission from Reference [66]. © 2012 Optical Society of America.

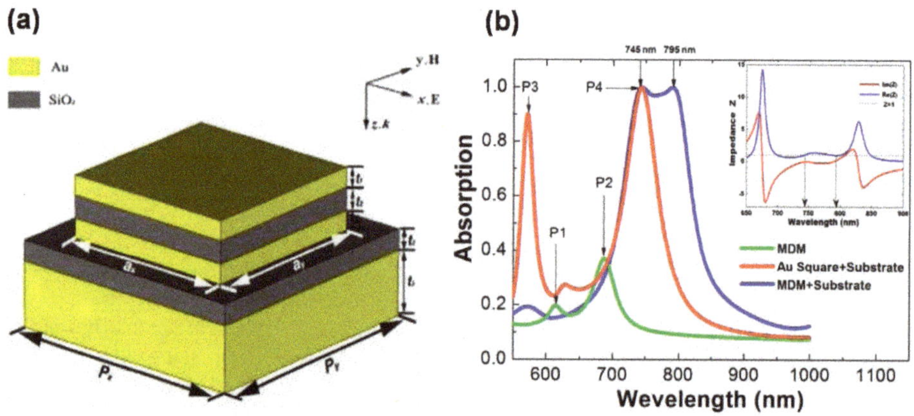

Figure 9. (**a**) Schematic illustration of a multi-layer MIM-PMA and (**b**) the calculated absorption spectrum. Figures are adapted with permission from Reference [95]. © 2012 Optical Society of America.

5. Advanced Structures and Applications

In this section, we briefly outline the advanced MIM-PMAs structures and applications other than absorbers to clarify the future research of MIM-PMAs. There has been growing interest in mainly three categories of advanced MIM-PMAs: flexible devices, the combination of graphene and other 2D materials, and metalenses.

One of the advanced structures is the flexible MIM-PMA [86,106,107], as shown in Figure 10 [86]. Flexible substrates such as Kapton film [86], polyethylene terephthalate (PET) [106] or polydimethylsiloxane (PDMS) [107] have enabled flexible MIM structures. MIM-PMAs coated on such flexible substrates can thus realize flexible and stretchable devices such as flexible solar cells, health care systems for the human body, and the cloaking of non-flat objects [108].

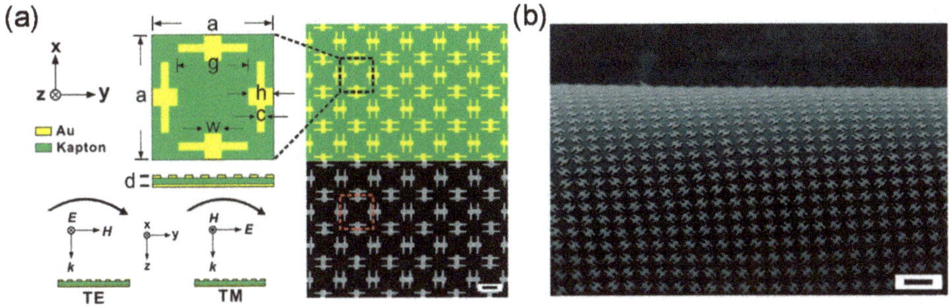

Figure 10. (**a**) Schematic illustration and (**b**) SEM image of a flexible MIM-PMA. Figures are adapted with permission from Reference [86]. © 2011 American Chemical Society.

The combination of graphene [109] and other 2D atomic layer materials [110] such as MoS_2 and WSe_2 with MIM-PMAs [55,81,111–115] is also drawing significant interest because these 2D atomic layer materials can strongly interact with plasmonic resonance [116]. Figure 11 shows a schematic illustration of graphene coated on an MIM-PMA (GMIM-PMA).

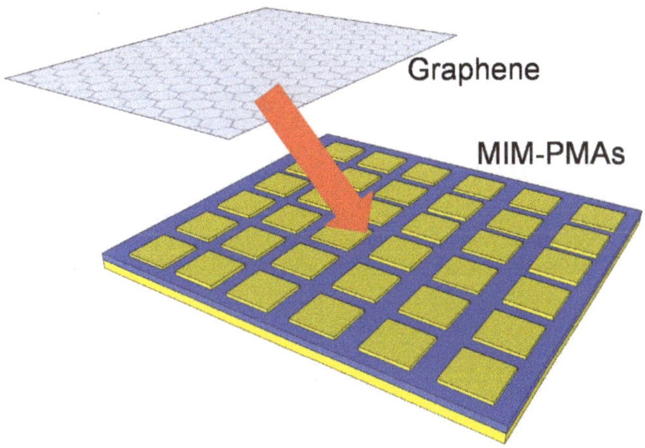

Figure 11. Schematic illustration of a graphene-coated MIM-PMA.

MIM-PMAs serve as a platform to enhance graphene absorption and realize high-performance graphene-based photodetectors [81,111,112,114]. The Fermi level of graphene can be electrically tuned according to the applied voltage; therefore, the absorption wavelength [117], reflection angle [118], and phase [119,120] can be electrically tuned by the applied voltage for graphene.

Metalenses are a new type of flat lens based on geometrical phase control [121–123]. MIM-PMAs are considered as an array of optical resonators that can introduce a desired spatial profile of the optical phase and consequently mold the wavefront [123]. Figure 12 shows a schematic illustration of a metalens or reflector using MIM-PMAs [123]. The MIM-PMA structures control the reflection and their phase by phase gradient surface structures with different sized micropatches on the planar surface. Strong plasmonic resonance can change the phase of the reflection and thereby realize geometrical control of the phase. As a result, a flat metalens can be realized.

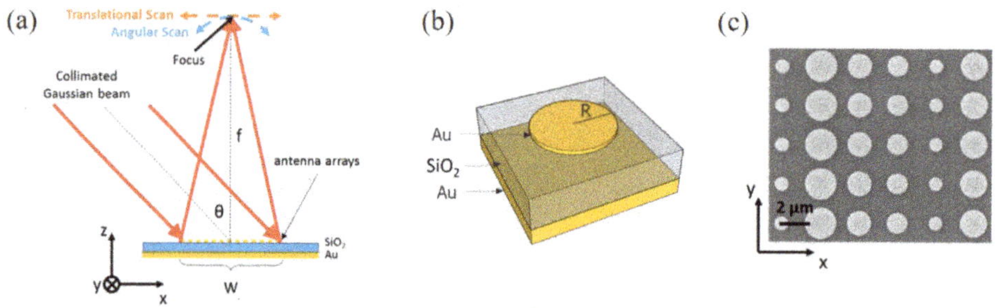

Figure 12. (**a**) Schematic illustration of a flat metalens using MIM-PMAs. (**b**) Schematic illustration of the unit cell of a reflector array lens. (**c**) SEM image of the metalens surface. Figures are adapted with permission from Reference [123]. © 2016 Optical Society of America.

There are other rapidly growing research fields of MIM-PMAs, such as the non-linear response of second [124,125] and third [126] harmonic generation, and reflection control [127]. As discussed in this section, although MIM-PMAs are simple structures, they have significant potential for novel physics and applications.

6. Conclusions

MIM-PMAs have been reviewed here in terms of their structures, basic principles of absorption, materials used, absorption properties of incident angle and polarization dependence, and strategies of multiband or broadband operation to clarify the design strategies for visible and IR wavelengths. The same principles can be applied for a wide range of wavelength regions such as the ultraviolet [128], terahertz [129–132], and microwave [133–135] regions.

A single MIM structure can be considered as a single optical antenna with strong LSPPs. Therefore, MIM structures are free from the restriction of periodicity and beyond the diffraction limit, making them able to realize strong absorption and geometrical tunability of the absorption wavelength with a much thinner and smaller absorber volume than conventional EM absorbers. Such controllability opens up a new stage of EM absorber research and many novel applications are expected.

In the future research of MIM-PMAs, flexible structures are also important to expand applications, such as health care devices for human sensing. The combination of new materials such as graphene and other 2D atomic materials with MIM-PMAs gives rise to the electrical tunability of the absorption wavelength, the phase, and the reflection angle because their optical constant can be tuned according to the applied voltage. MIM-PMAs can control other wavelengths than the absorption wavelength used. Thus, MIM-PMAs can be used for flat metalenses that can go beyond the diffraction limit.

We hope that this review paper will contribute to the development of advanced MM-PMAs and the expansion of their fields of application.

Conflicts of Interest: The authors declare no conflict of interest.

References

1. Ruck, G.T.; Barrick, D.E.; Stuart, W. *Radar Cross Section Handbook*; Peninsula Publishing: Los Altos, CA, USA, 2002; Volume 2, pp. 612–630.
2. Keasler, C.A.; Bellotti, E. A numerical study of broadband absorbers for visible to infrared detectors. *Appl. Phys. Lett.* **2011**, *99*. [CrossRef]
3. Li, P.; Liu, B.; Ni, Y.; Liew, K.K.; Sze, J.; Chen, S.; Shen, S. Large-Scale Nanophotonic Solar Selective Absorbers for High-Efficiency Solar Thermal Energy Conversion. *Adv. Mater.* **2015**, *27*, 4585–4591. [CrossRef] [PubMed]

4. Vorobyev, A.Y.; Topkov, A.N.; Gurin, O.V.; Svich, V.A.; Guo, C. Enhanced absorption of metals over ultrabroad electromagnetic spectrum. *Appl. Phys. Lett.* **2009**, *95*. [CrossRef]
5. Bae, K.; Kang, G.; Cho, S.K.; Park, W.; Kim, K.; Padilla, W.J. Flexible thin-film black gold membranes with ultrabroadband plasmonic nanofocusing for efficient solar vapour generation. *Nat. Commun.* **2015**, *6*. [CrossRef] [PubMed]
6. Munk, B.A. *Frequency Selective Surfaces: Theory and Design*; John Wiley & Sons: New York, NY, USA, 2000; pp. 315–325.
7. Salisbury, W.W. Absorbent Body for Electromagnetic Waves. U.S. Patent 2599944 A, 10 June 1952.
8. Knott, E.F.; Lunden, C.D. The two-sheet capacitive Jaumann absorber. *IEEE Trans. Antennas Propag.* **1995**, *43*, 1339–1343. [CrossRef]
9. Kawata, S. Plasmonics: Future outlook. *Jpn. J. Appl. Phys.* **2013**, *52*. [CrossRef]
10. Tanaka, T. Plasmonic metamaterials. *IEICE Electron. Express* **2012**, *9*, 34–50. [CrossRef]
11. Maier, S.A.; Atwater, H.A. Plasmonics: Localization and guiding of electromagnetic energy in metal/dielectric structures. *J. Appl. Phys.* **2005**, *98*. [CrossRef]
12. Gramotnev, D.K.; Bozhevolnyi, S.I. Plasmonics beyond the diffraction limit. *Nat. Photonics* **2010**, *4*, 83–91. [CrossRef]
13. Ogawa, S.; Okada, K.; Fukushima, N.; Kimata, M. Wavelength selective uncooled infrared sensor by plasmonics. *Appl. Phys. Lett.* **2012**, *100*. [CrossRef]
14. Ogawa, S.; Komoda, J.; Masuda, K.; Kimata, M. Wavelength selective wideband uncooled infrared sensor using a two-dimensional plasmonic absorber. *Opt. Eng.* **2013**, *52*. [CrossRef]
15. Takagawa, Y.; Ogawa, S.; Kimata, M. Detection wavelength control of uncooled infrared sensors by two-dimensional lattice plasmonic absorbers. *Sensors* **2015**, *15*, 13660–13669. [CrossRef] [PubMed]
16. Miyazaki, H.T.; Ikeda, K.; Kasaya, T.; Yamamoto, K.; Inoue, Y.; Fujimura, K.; Kanakugi, T.; Okada, M.; Hatade, K.; Kitagawa, S. Thermal emission of two-color polarized infrared waves from integrated plasmon cavities. *Appl. Phys. Lett.* **2008**, *92*. [CrossRef]
17. Ikeda, K.; Miyazaki, H.T.; Kasaya, T.; Yamamoto, K.; Inoue, Y.; Fujimura, K.; Kanakugi, T.; Okada, M.; Hatade, K.; Kitagawa, S. Controlled thermal emission of polarized infrared waves from arrayed plasmon nanocavities. *Appl. Phys. Lett.* **2008**, *92*. [CrossRef]
18. Ogawa, S.; Kimata, M. Direct fabrication and characterization of high-aspect-ratio plasmonic nanogratings using tapered-sidewall molds. *Opt. Mater. Express* **2017**, *7*, 633–640. [CrossRef]
19. Ogawa, S.; Takagawa, Y.; Kimata, M. Broadband polarization-selective uncooled infrared sensors using tapered plasmonic micrograting absorbers. *Sens. Actuators A* **2018**, *269*, 563–568. [CrossRef]
20. Lee, Y.P.; Rhee, J.Y.; Yoo, Y.J.; Kim, K.W. *Metamaterials for Perfect Absorption*; Springer: Singapore, 2016.
21. Mann, S.A.; Garnett, E.C. Resonant Nanophotonic Spectrum Splitting for Ultrathin Multijunction Solar Cells. *ACS Photonics* **2015**, *2*, 816–821. [CrossRef] [PubMed]
22. Liu, N.; Mesch, M.; Weiss, T.; Hentschel, M.; Giessen, H. Infrared Perfect Absorber and Its Application as Plasmonic Sensor. *Nano Lett.* **2010**, *10*, 2342–2348. [CrossRef] [PubMed]
23. Wang, G.; Chen, X.; Liu, S.; Wong, C.; Chu, S. Mechanical Chameleon through Dynamic Real-Time Plasmonic Tuning. *ACS Nano* **2016**, *10*, 1788–1794. [CrossRef] [PubMed]
24. Ni, X.; Wong, Z.J.; Mrejen, M.; Wang, Y.; Zhang, X. An ultrathin invisibility skin cloak for visible light. *Science* **2015**, *349*, 1310–1314. [CrossRef] [PubMed]
25. Qian, Z.; Kang, S.; Rajaram, V.; Cassella, C.; McGruer, N.E.; Rinaldi, M. Zero-power infrared digitizers based on plasmonically enhanced micromechanical photoswitches. *Nat. Nanotechnol.* **2017**, *12*, 969–973. [CrossRef] [PubMed]
26. Flauraud, V.; Reyes, M.; Paniagua-Domínguez, R.; Kuznetsov, A.I.; Brugger, J. Silicon Nanostructures for Bright Field Full Color Prints. *ACS Photonics* **2017**, *4*, 1913–1919. [CrossRef]
27. Miyata, M.; Hatada, H.; Takahara, J. Full-Color Subwavelength Printing with Gap-Plasmonic Optical Antennas. *Nano Lett.* **2016**, *16*, 3166–3172. [CrossRef] [PubMed]
28. Maier, T.; Brueckl, H. Multispectral microbolometers for the midinfrared. *Opt. Lett.* **2010**, *35*, 3766–3768. [CrossRef] [PubMed]
29. Chen, H.-H.; Su, Y.-C.; Huang, W.-L.; Kuo, C.-Y.; Tian, W.-C.; Chen, M.-J.; Lee, S.-C. A plasmonic infrared photodetector with narrow bandwidth absorption. *Appl. Phys. Lett.* **2014**, *105*. [CrossRef]

30. Ogawa, S.; Kimata, M. Wavelength- or Polarization-Selective Thermal Infrared Detectors for Multi-Color or Polarimetric Imaging Using Plasmonics and Metamaterials. *Materials* **2017**, *10*, 493. [CrossRef] [PubMed]
31. Suen, J.Y.; Fan, K.; Montoya, J.; Bingham, C.; Stenger, V.; Sriram, S.; Padilla, W.J. Multifunctional metamaterial pyroelectric infrared detectors. *Optica* **2017**, *4*, 276–279. [CrossRef]
32. Hui, Y.; Gomez-Diaz, J.S.; Qian, Z.; Alù, A.; Rinaldi, M. Plasmonic piezoelectric nanomechanical resonator for spectrally selective infrared sensing. *Nat. Commun.* **2016**, *7*. [CrossRef] [PubMed]
33. Ishikawa, A.; Tanaka, T. Metamaterial Absorbers for Infrared Detection of Molecular Self-Assembled Monolayers. *Sci. Rep.* **2015**, *5*. [CrossRef] [PubMed]
34. Aslan, E.; Aslan, E.; Turkmen, M.; Saracoglu, O.G. Metamaterial plasmonic absorber for reducing the spectral shift between near- and far-field responses in surface-enhanced spectroscopy applications. *Sens. Actuators A* **2017**, *267*, 60–69. [CrossRef]
35. Miyazaki, H.T.; Kasaya, T.; Iwanaga, M.; Choi, B.; Sugimoto, Y.; Sakoda, K. Dual-band infrared metasurface thermal emitter for CO_2 sensing. *Appl. Phys. Lett.* **2014**, *105*. [CrossRef]
36. Inoue, T.; De Zoysa, M.; Asano, T.; Noda, S. Realization of narrowband thermal emission with optical nanostructures. *Optica* **2015**, *2*, 27–35. [CrossRef]
37. Puscasu, I.; Schaich, W.L. Narrow-band, tunable infrared emission from arrays of microstrip patches. *Appl. Phys. Lett.* **2008**, *92*. [CrossRef]
38. Liu, X.; Starr, T.; Starr, A.F.; Padilla, W.J. Infrared spatial and frequency selective metamaterial with near-unity absorbance. *Phys. Rev. Lett.* **2010**, *104*. [CrossRef] [PubMed]
39. Wu, C.; Neuner, B.; Shvets, G.; John, J.; Milder, A.; Zollars, B.; Savoy, S. Large-area wide-angle spectrally selective plasmonic absorber. *Phys. Rev. B* **2011**, *84*. [CrossRef]
40. Chen, Y.B.; Chiu, F.C. Trapping mid-infrared rays in a lossy film with the Berreman mode, epsilon near zero mode, and magnetic polaritons. *Opt. Express* **2013**, *21*, 20771–20785. [CrossRef] [PubMed]
41. Hao, J.; Wang, J.; Liu, X.; Padilla, W.J.; Zhou, L.; Qiu, M. High performance optical absorber based on a plasmonic metamaterial. *Appl. Phys. Lett.* **2010**, *96*. [CrossRef]
42. Cheng, C.-W.; Abbas, M.N.; Chiu, C.-W.; Lai, K.-T.; Shih, M.-H.; Chang, Y.-C. Wide-angle polarization independent infrared broadband absorbers based on metallic multi-sized disk arrays. *Opt. Express* **2012**, *20*, 10376–10381. [CrossRef] [PubMed]
43. Hasan, D.; Pitchappa, P.; Wang, J.; Wang, T.; Yang, B.; Ho, C.P.; Lee, C. Novel CMOS-Compatible Mo–AlN–Mo Platform for Metamaterial-Based Mid-IR Absorber. *ACS Photonics* **2017**, *4*, 302–315. [CrossRef]
44. Monti, A.; Toscano, A.; Bilotti, F. Exploiting the surface dispersion of nanoparticles to design optical-resistive sheets and Salisbury absorbers. *Opt. Lett.* **2016**, *41*, 3383–3386. [CrossRef] [PubMed]
45. Dereshgi, S.A.; Okyay, A.K. Large area compatible broadband superabsorber surfaces in the VIS-NIR spectrum utilizing metal-insulator-metal stack and plasmonic nanoparticles. *Opt. Express* **2016**, *24*, 17644–17653. [CrossRef] [PubMed]
46. Monti, A.; Alù, A.; Toscano, A.; Bilotti, F. Narrowband transparent absorbers based on ellipsoidal nanoparticles. *Appl. Opt.* **2017**, *56*, 7533–7538. [CrossRef] [PubMed]
47. Zhang, B.; Zhao, Y.; Hao, Q.; Kiraly, B.; Khoo, I.-C.; Chen, S.; Huang, T.J. Polarization-independent dual-band infrared perfect absorber based on a metal-dielectric-metal elliptical nanodisk array. *Opt. Express* **2011**, *19*, 15221–15228. [CrossRef] [PubMed]
48. Chevalier, P.; Bouchon, P.; Jaeck, J.; Lauwick, D.; Bardou, N.; Kattnig, A.; Pardo, F.; Haïdar, R. Absorbing metasurface created by diffractionless disordered arrays of nanoantennas. *Appl. Phys. Lett.* **2015**, *107*. [CrossRef]
49. Aydin, K.; Ferry, V.E.; Briggs, R.M.; Atwater, H.A. Broadband polarization-independent resonant light absorption using ultrathin plasmonic super absorbers. *Nat. Commun.* **2011**, *2*, 517. [CrossRef] [PubMed]
50. Lin, L.; Zheng, Y. Optimizing plasmonic nanoantennas via coordinated multiple coupling. *Sci. Rep.* **2015**, *5*. [CrossRef] [PubMed]
51. Alici, K.B.; Turhan, A.B.; Soukoulis, C.M.; Ozbay, E. Optically thin composite resonant absorber at the near-infrared band: A polarization independent and spectrally broadband configuration. *Opt. Express* **2011**, *19*, 14260–14267. [CrossRef] [PubMed]
52. Chen, K.; Adato, R.; Altug, H. Dual-Band Perfect Absorber for Multispectral Plasmon-Enhanced Infrared Spectroscopy. *ACS Nano* **2012**, *6*, 7998–8006. [CrossRef] [PubMed]

53. Chen, N.; Pitchappa, P.; Ho, C.P.; Hasan, D.; Kropelnicki, P.; Alioto, M.; Lee, C. Polarization controllable multispectral symmetry-breaking absorber in mid-infrared. *J. Appl. Phys.* **2016**, *120*. [CrossRef]
54. Akselrod, G.M.; Huang, J.; Hoang, T.B.; Bowen, P.T.; Su, L.; Smith, D.R.; Mikkelsen, M.H. Large-Area Metasurface Perfect Absorbers from Visible to Near-Infrared. *Adv. Mater.* **2015**, *27*, 8028–8034. [CrossRef] [PubMed]
55. Huang, J.; Akselrod, G.M.; Ming, T.; Kong, J.; Mikkelsen, M.H. Tailored Emission Spectrum of 2D Semiconductors Using Plasmonic Nanocavities. *ACS Photonics* **2017**. [CrossRef]
56. Zhang, B.; Hendrickson, J.; Guo, J. Multispectral near-perfect metamaterial absorbers using spatially multiplexed plasmon resonance metal square structures. *J. Opt. Soc. Am. B* **2013**, *30*, 656–662. [CrossRef]
57. Liu, T.; Takahara, J. Ultrabroadband absorber based on single-sized embedded metal-dielectric-metal structures and application of radiative cooling. *Opt. Express* **2017**, *25*, A612–A627. [CrossRef] [PubMed]
58. Lefebvre, A.; Costantini, D.; Doyen, I.; Lévesque, Q.; Lorent, E.; Jacolin, D.; Greffet, J.J.; Boutami, S.; Benisty, H. CMOS compatible metal-insulator-metal plasmonic perfect absorbers. *Opt. Mater. Express* **2016**, *6*, 2389–2396. [CrossRef]
59. Li, W.; Guler, U.; Kinsey, N.; Naik, G.V.; Boltasseva, A.; Guan, J.; Shalaev, V.M.; Kildishev, A.V. Refractory Plasmonics with Titanium Nitride: Broadband Metamaterial Absorber. *Adv. Mater.* **2014**, *26*, 7959–7965. [CrossRef] [PubMed]
60. Wang, H.; Chen, Q.; Wen, L.; Song, S.; Hu, X.; Xu, G. Titanium-nitride-based integrated plasmonic absorber/emitter for solar thermophotovoltaic application. *Photonics Res.* **2015**, *3*, 329–334. [CrossRef]
61. Gorgulu, K.; Gok, A.; Yilmaz, M.; Topalli, K.; Bıyıklı, N.; Okyay, A.K. All-Silicon Ultra-Broadband Infrared Light Absorbers. *Sci. Rep.* **2016**, *6*. [CrossRef] [PubMed]
62. Thongrattanasiri, S.; Koppens, F.H.L.; García de Abajo, F.J. Complete Optical Absorption in Periodically Patterned Graphene. *Phys. Rev. Lett.* **2012**, *108*. [CrossRef] [PubMed]
63. Matsuno, Y.; Sakurai, A. Perfect infrared absorber and emitter based on a large-area metasurface. *Opt. Mater. Express* **2017**, *7*, 618–626. [CrossRef]
64. Du, K.; Li, Q.; Zhang, W.; Yang, Y.; Qiu, M. Wavelength and Thermal Distribution Selectable Microbolometers Based on Metamaterial Absorbers. *IEEE Photonics J.* **2015**, *7*. [CrossRef]
65. Cui, Y.; Xu, J.; Fung, K.H.; Jin, Y.; Kumar, A.; He, S.; Fang, N.X. A thin film broadband absorber based on multi-sized nanoantennas. *Appl. Phys. Lett.* **2011**, *99*. [CrossRef]
66. Bouchon, P.; Koechlin, C.; Pardo, F.; Haïdar, R.; Pelouard, J.-L. Wideband omnidirectional infrared absorber with a patchwork of plasmonic nanoantennas. *Opt. Lett.* **2012**, *37*, 1038–1040. [CrossRef] [PubMed]
67. Tittl, A.; Michel, A.K.; Schaferling, M.; Yin, X.; Gholipour, B.; Cui, L.; Wuttig, M.; Taubner, T.; Neubrech, F.; Giessen, H. A Switchable Mid-Infrared Plasmonic Perfect Absorber with Multispectral Thermal Imaging Capability. *Adv. Mater.* **2015**, *27*, 4597–4603. [CrossRef] [PubMed]
68. Wang, H.; Yang, Y.; Wang, L. Switchable wavelength-selective and diffuse metamaterial absorber/emitter with a phase transition spacer layer. *Appl. Phys. Lett.* **2014**, *105*. [CrossRef]
69. Ogawa, S.; Fujisawa, D.; Hata, H.; Uetsuki, M.; Misaki, K.; Kimata, M. Mushroom plasmonic metamaterial infrared absorbers. *Appl. Phys. Lett.* **2015**, *106*. [CrossRef]
70. Ogawa, S.; Fujisawa, D.; Hata, H.; Kimata, M. Absorption properties of simply fabricated all-metal mushroom plasmonic metamaterials incorporating tube-shaped posts for multi-color uncooled infrared image sensor applications. *Photonics* **2016**, *3*, 9. [CrossRef]
71. Ogawa, S.; Fujisawa, D.; Kimata, M. Theoretical investigation of all-metal based mushroom plasmonic metamaterial absorbers at infrared wavelengths. *Opt. Eng.* **2015**, *54*. [CrossRef]
72. Cheng, F.; Gao, J.; Luk, T.S.; Yang, X. Structural color printing based on plasmonic metasurfaces of perfect light absorption. *Sci. Rep.* **2015**, *5*. [CrossRef] [PubMed]
73. Bouchon, P.; Pardo, F.; Portier, B.; Ferlazzo, L.; Ghenuche, P.; Dagher, G.; Dupuis, C.; Bardou, N.; Haïdar, R.; Pelouard, J.-L. Total funneling of light in high aspect ratio plasmonic nanoresonators. *Appl. Phys. Lett.* **2011**, *98*. [CrossRef]
74. Koechlin, C.; Bouchon, P.; Pardo, F.; Pelouard, J.-L.; Haïdar, R. Analytical description of subwavelength plasmonic MIM resonators and of their combination. *Opt. Express* **2013**, *21*, 7025–7032. [CrossRef] [PubMed]
75. Zhou, W.; Li, K.; Song, C.; Hao, P.; Chi, M.; Yu, M.; Wu, Y. Polarization-independent and omnidirectional nearly perfect absorber with ultra-thin 2D subwavelength metal grating in the visible region. *Opt. Express* **2015**, *23*, A413–A418. [CrossRef] [PubMed]

76. Ra'di, Y.; Asadchy, V.S.; Kosulnikov, S.U.; Omelyanovich, M.M.; Morits, D.; Osipov, A.V.; Simovski, C.R.; Tretyakov, S.A. Full Light Absorption in Single Arrays of Spherical Nanoparticles. *ACS Photonics* **2015**, *2*, 653–660. [CrossRef]
77. Li, Z.; Butun, S.; Aydin, K. Large-Area, Lithography-Free Super Absorbers and Color Filters at Visible Frequencies Using Ultrathin Metallic Films. *ACS Photonics* **2015**, *2*, 183–188. [CrossRef]
78. Ding, F.; Mo, L.; Zhu, J.; He, S. Lithography-free, broadband, omnidirectional, and polarization-insensitive thin optical absorber. *Appl. Phys. Lett.* **2015**, *106*. [CrossRef]
79. Peng, H.; Luo, Y.; Ying, X.; Pu, Y.; Jiang, Y.; Xu, J.; Liu, Z. Broadband and highly absorbing multilayer structure in mid-infrared. *Appl. Opt.* **2016**, *55*, 8833–8838. [CrossRef] [PubMed]
80. Chettiar, U.K.; Kildishev, A.V.; Klar, T.A.; Shalaev, V.M. Negative index metamaterial combining magnetic resonators with metal films. *Opt. Express* **2006**, *14*, 7872–7877. [CrossRef] [PubMed]
81. Ogawa, S.; Fujisawa, D.; Shimatani, M.; Matsumoto, K. Graphene on plasmonic metamaterials for infrared detection. In *Infrared Technology and Applications XLII*; Andresen, B.F., Fulop, G.F., Hanson, C.M., Miller, J.L., Norton, P.R., Eds.; SPIE: Bellingham, WA, USA, 2016; Volume 9819, p. 98191S.
82. Tyo, J.S.; Goldstein, D.L.; Chenault, D.B.; Shaw, J.A. Review of passive imaging polarimetry for remote sensing applications. *Appl. Opt.* **2006**, *45*, 5453–5469. [CrossRef] [PubMed]
83. Schott, J.R. *Fundamentals of Polarimetric Remote Sensing*; SPIE: Bellingham, WA, USA, 2009.
84. Gurton, K.P.; Yuffa, A.J.; Videen, G.W. Enhanced facial recognition for thermal imagery using polarimetric imaging. *Opt. Lett.* **2014**, *39*, 3857–3859. [CrossRef] [PubMed]
85. Chen, S.; Cheng, H.; Yang, H.; Li, J.; Duan, X.; Gu, C.; Tian, J. Polarization insensitive and omnidirectional broadband near perfect planar metamaterial absorber in the near infrared regime. *Appl. Phys. Lett.* **2011**, *99*. [CrossRef]
86. Jiang, Z.H.; Yun, S.; Toor, F.; Werner, D.H.; Mayer, T.S. Conformal Dual-Band Near-Perfectly Absorbing Mid-Infrared Metamaterial Coating. *ACS Nano* **2011**, *5*, 4641–4647. [CrossRef] [PubMed]
87. Butun, S.; Aydin, K. Structurally tunable resonant absorption bands in ultrathin broadband plasmonic absorbers. *Opt. Express* **2014**, *22*, 19457–19468. [CrossRef] [PubMed]
88. Bai, Y.; Zhao, L.; Ju, D.; Jiang, Y.; Liu, L. Wide-angle, polarization-independent and dual-band infrared perfect absorber based on L-shaped metamaterial. *Opt. Express* **2015**, *23*, 8670–8680. [CrossRef] [PubMed]
89. Ghobadi, A.; Hajian, H.; Gokbayrak, M.; Dereshgi, S.A.; Toprak, A.; Butun, B.; Ozbay, E. Visible light nearly perfect absorber: An optimum unit cell arrangement for near absolute polarization insensitivity. *Opt. Express* **2017**, *25*, 27624–27634. [CrossRef] [PubMed]
90. Koechlin, C.; Bouchon, P.; Pardo, F.; Jaeck, J.; Lafosse, X.; Pelouard, J.-L.; Haïdar, R. Total routing and absorption of photons in dual color plasmonic antennas. *Appl. Phys. Lett.* **2011**, *99*. [CrossRef]
91. Bossard, J.A.; Lin, L.; Yun, S.; Liu, L.; Werner, D.H.; Mayer, T.S. Near-Ideal Optical Metamaterial Absorbers with Super-Octave Bandwidth. *ACS Nano* **2014**, *8*, 1517–1524. [CrossRef] [PubMed]
92. Xu, J.; Zhao, Z.; Yu, H.; Yang, L.; Gou, P.; Cao, J.; Zou, Y.; Qian, J.; Shi, T.; Ren, Q.; An, Z. Design of triple-band metamaterial absorbers with refractive index sensitivity at infrared frequencies. *Opt. Express* **2016**, *24*, 25742–25751. [CrossRef] [PubMed]
93. Khan, A.D.; Amin, M. Tunable Salisbury Screen Absorber Using Square Lattice of Plasmonic Nanodisk. *Plasmonics* **2016**, *12*, 257–262. [CrossRef]
94. Kim, J.; Han, K.; Hahn, J.W. Selective dual-band metamaterial perfect absorber for infrared stealth technology. *Sci. Rep.* **2017**, *7*. [CrossRef] [PubMed]
95. Wang, J.; Fan, C.; Ding, P.; He, J.; Cheng, Y.; Hu, W.; Cai, G.; Liang, E.; Xue, Q. Tunable broad-band perfect absorber by exciting of multiple plasmon resonances at optical frequency. *Opt. Express* **2012**, *20*, 14871–14878. [CrossRef] [PubMed]
96. Cui, Y.; Fung, K.H.; Xu, J.; Ma, H.; Jin, Y.; He, S.; Fang, N.X. Ultrabroadband Light Absorption by a Sawtooth Anisotropic Metamaterial Slab. *Nano Lett.* **2012**, *12*, 1443–1447. [CrossRef] [PubMed]
97. Lobet, M.; Lard, M.; Sarrazin, M.; Deparis, O.; Henrard, L. Plasmon hybridization in pyramidal metamaterials: A route towards ultra-broadband absorption. *Opt. Express* **2014**, *22*, 12678–12690. [CrossRef] [PubMed]
98. Sakurai, A.; Zhao, B.; Zhang, Z.M. Resonant frequency and bandwidth of metamaterial emitters and absorbers predicted by an RLC circuit model. *J. Quant. Spectrosc. Radiant. Transf.* **2014**, *149*, 33–40. [CrossRef]

99. Adomanis, B.M.; Watts, C.M.; Koirala, M.; Liu, X.; Tyler, T.; West, K.G.; Starr, T.; Bringuier, J.N.; Starr, A.F.; Jokerst, N.M.; et al. Bi-layer metamaterials as fully functional near-perfect infrared absorbers. *Appl. Phys. Lett.* **2015**, *107*. [CrossRef]
100. Wu, D.; Liu, C.; Liu, Y.; Yu, L.; Yu, Z.; Chen, L.; Ma, R.; Ye, H. Numerical study of an ultra-broadband near-perfect solar absorber in the visible and near-infrared region. *Opt. Lett.* **2017**, *42*, 450–453. [CrossRef] [PubMed]
101. Üstün, K.; Turhan-Sayan, G. Broadband LWIR and MWIR metamaterial absorbers with a simple design topology: Almost perfect absorption and super-octave band operation in MWIR band. *J. Opt. Soc. Am. B* **2017**, *34*, D86–D94. [CrossRef]
102. Hubarevich, A.; Kukhta, A.; Demir, H.V.; Sun, X.; Wang, H. Ultra-thin broadband nanostructured insulator-metal-insulator-metal plasmonic light absorber. *Opt. Express* **2015**, *23*, 9753–9761. [CrossRef] [PubMed]
103. Wu, C.; Shvets, G. Design of metamaterial surfaces with broadband absorbance. *Opt. Lett.* **2012**, *37*, 308–310. [CrossRef] [PubMed]
104. Harris, L.; McGinnies, R.T.; Siegel, B.M. The Preparation and Optical Properties of Gold Blacks. *J. Opt. Soc. Am.* **1948**, *38*, 582–589. [CrossRef]
105. Harris, L. The Transmittance and Reflectance of Gold Black Deposits in the 15- to 100-Micron Region. *J. Opt. Soc. Am.* **1961**, *51*, 80–82. [CrossRef]
106. Li, G.X.; Chen, S.M.; Wong, W.H.; Pun, E.Y.B.; Cheah, K.W. Highly flexible near-infrared metamaterials. *Opt. Express* **2012**, *20*, 397–402. [CrossRef] [PubMed]
107. Ji, T.; Wang, Y.; Cui, Y.; Lin, Y.; Hao, Y.; Li, D. Flexible broadband plasmonic absorber on moth-eye substrate. *Mater. Today Energy* **2017**, *5*, 181–186. [CrossRef]
108. Walia, S.; Shah, C.M.; Gutruf, P.; Nili, H.; Chowdhury, D.R.; Withayachumnankul, W.; Bhaskaran, M.; Sriram, S. Flexible metasurfaces and metamaterials: A review of materials and fabrication processes at micro- and nano-scales. *Appl. Phys. Rev.* **2015**, *2*. [CrossRef]
109. Novoselov, K.S.; Geim, A.K.; Morozov, S.V.; Jiang, D.; Zhang, Y.; Dubonos, S.V.; Grigorieva, I.V.; Firsov, A.A. Electric field effect in atomically thin carbon films. *Science* **2004**, *306*, 666–669. [CrossRef] [PubMed]
110. Sun, Z.; Chang, H. Graphene and Graphene-like Two-Dimensional Materials in Photodetection: Mechanisms and Methodology. *ACS Nano* **2014**, *8*, 4133–4156. [CrossRef] [PubMed]
111. Cai, Y.; Zhu, J.; Liu, Q.H. Tunable enhanced optical absorption of graphene using plasmonic perfect absorbers. *Appl. Phys. Lett.* **2015**, *106*. [CrossRef]
112. Pan, Q.; Hong, J.; Zhang, G.; Shuai, Y.; Tan, H. Graphene plasmonics for surface enhancement near-infrared absorptivity. *Opt. Express* **2017**, *25*, 16400–16408. [CrossRef] [PubMed]
113. Jariwala, D.; Davoyan, A.R.; Tagliabue, G.; Sherrott, M.C.; Wong, J.; Atwater, H.A. Near-Unity Absorption in van der Waals Semiconductors for Ultrathin Optoelectronics. *Nano Lett.* **2016**, *16*, 5482–5487. [CrossRef] [PubMed]
114. Ogawa, S.; Shimatani, M.; Fukushima, S.; Okuda, S.; Matsumoto, K. Graphene on metal-insulator-metal-based plasmonic metamaterials at infrared wavelengths. *Opt. Express* **2018**, *26*. [CrossRef] [PubMed]
115. Huang, H.; Xia, H.; Xie, W.; Guo, Z.; Li, H.; Xie, D. Design of broadband graphene-metamaterial absorbers for permittivity sensing at mid-infrared regions. *Sci. Rep.* **2018**, *8*. [CrossRef] [PubMed]
116. Low, T.; Chaves, A.; Caldwell, J.D.; Kumar, A.; Fang, N.X.; Avouris, P.; Heinz, T.F.; Guinea, F.; Martin-Moreno, L.; Koppens, F. Polaritons in layered two-dimensional materials. *Nat. Mater.* **2017**, *16*, 182–194. [CrossRef] [PubMed]
117. Yao, Y.; Shankar, R.; Kats, M.A.; Song, Y.; Kong, J.; Loncar, M.; Capasso, F. Electrically tunable metasurface perfect absorbers for ultrathin mid-infrared optical modulators. *Nano Lett.* **2014**, *14*, 6526–6532. [CrossRef] [PubMed]
118. Su, X.; Wei, Z.; Wu, C.; Long, Y.; Li, H. Negative reflection from metal/graphene plasmonic gratings. *Opt. Lett.* **2016**, *41*, 348–351. [CrossRef] [PubMed]
119. Miao, Z.; Wu, Q.; Li, X.; He, Q.; Ding, K.; An, Z.; Zhang, Y.; Zhou, L. Widely Tunable Terahertz Phase Modulation with Gate-Controlled Graphene Metasurfaces. *Phys. Rev. X* **2015**, *5*, 041027. [CrossRef]
120. Sherrott, M.C.; Hon, P.W.C.; Fountaine, K.T.; Garcia, J.C.; Ponti, S.M.; Brar, V.W.; Sweatlock, L.A.; Atwater, H.A. Experimental Demonstration of >230° Phase Modulation in Gate-Tunable Graphene–Gold Reconfigurable Mid-Infrared Metasurfaces. *Nano Lett.* **2017**, *17*, 3027–3034. [CrossRef] [PubMed]

121. Aieta, F.; Kats, M.A.; Genevet, P.; Capasso, F. Multiwavelength achromatic metasurfaces by dispersive phase compensation. *Science* **2015**, *347*, 1342–1345. [CrossRef] [PubMed]
122. Khorasaninejad, M.; Chen, W.T.; Zhu, A.Y.; Oh, J.; Devlin, R.C.; Roques-Carmes, C.; Mishra, I.; Capasso, F. Visible Wavelength Planar Metalenses Based on Titanium Dioxide. *IEEE J. Sel. Top. Quantum Electron.* **2017**, *23*. [CrossRef]
123. Zhang, S.; Kim, M.-H.; Aieta, F.; She, A.; Mansuripur, T.; Gabay, I.; Khorasaninejad, M.; Rousso, D.; Wang, X.; Troccoli, M.; et al. High efficiency near diffraction-limited mid-infrared flat lenses based on metasurface reflectarrays. *Opt. Express* **2016**, *24*, 18024–18034. [CrossRef] [PubMed]
124. Lee, J.; Tymchenko, M.; Argyropoulos, C.; Chen, P.-Y.; Lu, F.; Demmerle, F.; Boehm, G.; Amann, M.-C.; Alù, A.; Belkin, M.A. Giant nonlinear response from plasmonic metasurfaces coupled to intersubband transitions. *Nature* **2014**, *511*, 65–69. [CrossRef] [PubMed]
125. Lu, X.; Zhang, L.; Zhang, T. Nanoslit-microcavity-based narrow band absorber for sensing applications. *Opt. Express* **2015**, *23*, 20715–20720. [CrossRef] [PubMed]
126. Lassiter, J.B.; Chen, X.; Liu, X.; Ciracì, C.; Hoang, T.B.; Larouche, S.; Oh, S.-H.; Mikkelsen, M.H.; Smith, D.R. Third-Harmonic Generation Enhancement by Film-Coupled Plasmonic Stripe Resonators. *ACS Photonics* **2014**, *1*, 1212–1217. [CrossRef]
127. Huang, Y.-W.; Lee, H.W.H.; Sokhoyan, R.; Pala, R.A.; Thyagarajan, K.; Han, S.; Tsai, D.P.; Atwater, H.A. Gate-Tunable Conducting Oxide Metasurfaces. *Nano Lett.* **2016**, *16*, 5319–5325. [CrossRef] [PubMed]
128. Sabah, C.; Esat, V.; Obaidullah, M. Polarization independent triple-band (5,4) semiconducting carbon nanotube metamaterial absorber design for visible and ultraviolet regions. *J. Nanophotonics* **2017**, *11*, 046011. [CrossRef]
129. Diem, M.; Koschny, T.; Soukoulis, C. Wide-angle perfect absorber/thermal emitter in the terahertz regime. *Phys. Rev. B* **2009**, *79*. [CrossRef]
130. Tao, H.; Bingham, C.; Strikwerda, A.; Pilon, D.; Shrekenhamer, D.; Landy, N.; Fan, K.; Zhang, X.; Padilla, W.; Averitt, R. Highly flexible wide angle of incidence terahertz metamaterial absorber: Design, fabrication, and characterization. *Phys. Rev. B* **2008**, *78*. [CrossRef]
131. Alves, F.; Grbovic, D.; Karunasiri, G. Investigation of microelectromechanical systems bimaterial sensors with metamaterial absorbers for terahertz imaging. *Opt. Eng.* **2014**, *53*. [CrossRef]
132. Yahiaoui, R.; Hanai, K.; Takano, K.; Nishida, T.; Miyamaru, F.; Nakajima, M.; Hangyo, M. Trapping waves with terahertz metamaterial absorber based on isotropic Mie resonators. *Opt. Lett.* **2015**, *40*, 3197–3200. [CrossRef] [PubMed]
133. Landy, N.; Sajuyigbe, S.; Mock, J.; Smith, D.; Padilla, W. Perfect Metamaterial Absorber. *Phys. Rev. Lett.* **2008**, *100*. [CrossRef] [PubMed]
134. Zhou, W.; Wang, P.; Wang, N.; Jiang, W.; Dong, X.; Hu, S. Microwave metamaterial absorber based on multiple square ring structures. *AIP Adv.* **2015**, *5*. [CrossRef]
135. Ding, F.; Cui, Y.; Ge, X.; Jin, Y.; He, S. Ultra-broadband microwave metamaterial absorber. *Appl. Phys. Lett.* **2012**, *100*, 103506. [CrossRef]

materials

MDPI

Article

Soft and Stiff Simplex Tensegrity Lattices as Extreme Smart Metamaterials

Anna Al Sabouni-Zawadzka * and Wojciech Gilewski

Faculty of Civil Engineering, Warsaw University of Technology, 00-637 Warsaw, Poland; w.gilewski@il.pw.edu.pl
* Correspondence: a.sabouni@il.pw.edu.pl; Tel.: +48-22-234-5753

Received: 9 November 2018; Accepted: 3 January 2019; Published: 8 January 2019

Abstract: The present paper is dedicated to an evaluation of novel cellular metamaterials based on a tensegrity pattern. The materials are constructed from supercells, each of which consists of a number of simplex modules with different geometrical proportions. Mechanical properties of the metamaterial can be controlled by adjusting the level of self-equilibrated forces or by changing the properties of structural members. A continuum model based on the equivalence of strain energy of the 3D theory of elasticity with a discrete formulation is used to identify the qualitative properties of the considered metamaterials. The model allows the inclusion of nonlinearities related to the equations of equilibrium in actual configuration of the structure with self-equilibrated set of normal forces typical for tensegrities. The lattices are recognised as extreme metamaterials according to the eigensolution of the equivalent elasticity matrices of the continuum model. The six representative deformation modes are defined and discussed: stiff, soft and medium extensional modes and high (double) as well as low shear modes. The lattices are identified as unimode or nearly bimode according to the classification of extreme materials.

Keywords: metamaterial; tensegrity lattice; extreme material

1. Introduction

Identification of unusual mechanical properties of tensegrity structures is one of the interesting challenges of exploring the mechanics of engineering metamaterials. Metamaterials are usually defined as man-made composites that do not exist in nature and have atypical or unusual properties [1,2]. In recent years, there have been many significant and important scientific studies in the field of metamaterials with unusual mechanical properties [2] such as: negative Poisson's ratio, atypical dynamic characteristics, unusual volume changes and expansion modules, very light materials, ultra-stiff or ultra-soft materials. Smart materials are defined as those in which one form of energy (mechanical, magnetic, electrical, etc.) is transformed into another one in a reversible and repeatable way [3]. Smart structures are capable of diagnosing changes in the environment and reacting to them in an adaptive way [4]. These features distinguish them from typical structures, whose main purpose is to provide load capacity and to ensure the safety of people. At the same time, we observe the adaptability to functional needs, modifications of the shape of the structure, changes of stiffness or damping properties, in order to minimize deformation and possible damage. Taking into account the above definitions, a smart metamaterial should be regarded as a smart structure rather than as a smart material.

An interesting idea of modern research in this field is metamaterials based on origami patterns [5,6], which are inspired by the ancient art of paper folding. One of the most popular and effective patterns are the Miura-Ori folds. Similarly to the origami patterns, tensegrities seem to show a great potential for construction of metamaterials with non-typical mechanical characteristics [7–9]. For the purpose of the present work, tensegrities are defined as cable-strut systems with a special configuration of nodes. They are statically indeterminate structures in a stable equilibrium. Struts form a discontinuous system of members under compression that is surrounded by a continuous system of elements in tension, which exhibit no compressive stiffness. Particular features of tensegrity structures are infinitesimal mechanisms, balanced with self-stress states [10]. Occurrence of a self-stress state in a structure indicates that there is a certain set of internal forces in structural members, which are independent from external loading and boundary conditions because they are in self-equilibrium.

The main advantages of pin-joined tensegrity structures are: high stiffness-to-mass ratio, controllability, reliability and deployability [11,12]. Furthermore, tensegrities exhibit some special features, which are a result of the occurrence of infinitesimal mechanisms that are stabilized by the system of normal forces in self-equilibrium. One can control their static and dynamic characteristics with an adjustment of the pre-stressing forces [13,14].

As was presented in reference [11], there are some particular features of tensegrity structures following which one can classify them as smart structures. These features are: self-control, self-diagnosis, self-repair and self-adjustment (active control) with the use of self-stress as well as geometrical properties of the structure.

The term "tensegrity metamaterial" was first introduced in reference [15] for the dynamics of the chain of tensegrity prisms and developed in references [16,17]. A similar concept was analyzed in reference [18]. Self-similar tensegrity columns of order 1 and higher order were proposed in reference [19]. The paper reference [20] is dedicated to the morphological optimization of tensegrity-type metamaterials with a prototypical model of an infinite slab. The formulation of the novel class $\theta = 1$ units was presented in reference [21] with the discussion of potential of such structures for mechanical metamaterials. Geometrically nonlinear behavior of uniformly compressed tensegrity prisms with extreme softening/stiffening response is recognized in reference [22] with relation to the design and manufacture of tensegrity lattices and innovative metamaterials. An interesting method to construct 3D tensegrity lattices from truncated octahedron elementary cells was proposed and discussed in reference [23] and extended for phase transition in reference [24]. Various automatically assembled tensegrity lattices were proposed in reference [25] for large scale structures. Metal rubber was introduced into the struts of a tensegrity prisms in references [26,27]. Both the theoretical and experimental data show the significant improvement of energy absorption and tunable dynamic properties to create an efficient mechanical metamaterial. Smart properties as well as a negative Poisson's ratio were observed in reference [28] for the orthotropic metamaterial based on the simplex tensegrity pattern. To conclude, great potential and dynamic development of the tensegrity based metamaterials have been observed in the literature of the last few years.

The concept of extreme materials was introduced in reference [29]. They are defined as extremely stiff materials under the action of certain stresses, or extremely compliant in other orthogonal cases of stresses. Extreme materials often have negative Poisson's ratios. The study of extreme properties of materials is based on the analysis of elasticity tensor. As is known, it must be positive definite and in the theory of elasticity it shows certain types of symmetries. This tensor can be diagonalized by orthogonal transformation. If we present the components of the elastic tensor in the Voight's form as a square matrix $\mathbf{E}$ of dimensions 6×6, its diagonal representation is the set of eigenvalues $\lambda_i > 0$ ($i = 1, 2, \dots, 6$), and orthogonal eigenvectors $\mathbf{w}_i$ describe the appropriate forms of deformation. One can classify materials as nullmode, unimode, bimode, trimode, quadramode, pentamode or hexamode [29,30] depending on the number of eigenvalues λ_i that are very small. Such a classification is often used in determining the properties of metamaterials, as long as the elastic matrix $\mathbf{E}$ is known.

The present paper is dedicated to the analysis of the tensegrity inspired metamaterials in view of possible extreme properties. The continuum model of mechanical properties of tensegrity lattices is used to define the equivalent elasticity matrix $\mathbf{E}$. The proposed continuum model [28,31] is applied to identify the extreme properties of the proposed metamaterial. Its extreme characteristics can be controlled with the self-stress state and cable to strut properties ratio, following the features of smart metamaterials. Three representative tensegrity lattices based on the 4-strut simplex module are discussed. According to the best knowledge of the authors there are no papers in this field in the available literature.

2. Continuum Model of a Tensegrity Lattice

The continuum model is based on the comparison of the strain energy of a tensegrity structure defined using a discrete model and the strain energy of a solid determined according to the symmetric 3D elasticity theory [28,31].

A discrete model describes a tensegrity pin-joined truss structure which is composed of e straight and prismatic bars of the lengths l_k, cross sections A_k and Young's modulus E_k. The bars are connected in nodes in which a number of s nodal displacements q_j and nodal forces Q_j are defined (see [32,33] for details). Axial forces N_k can be expressed by the extensions of bars Δ_k in the form $N_k = E_k A_k \Delta_k / l_k$. The extensions Δ_k are a combination of nodal displacements $\Delta_k = \sum_{j=1}^{s} B_{kj} q_j$, $J = 1,2, \ldots, s$. B_{kj} is a compatibility matrix of projection of nodal displacements for the directions of bar axes [32,33]. Additionally, the self-equilibrated system of axial forces S_k, which satisfy the homogeneous set of equilibrium equations $\sum_{k=1}^{e} B_{jk} S_k = 0$, is considered. If one considers the equations of equilibrium in the actual configuration, the moment $M_k = S_k l_k \psi_k$ is acting on each bar. Angles of bar rotations ψ_k can be expressed as a combination of nodal displacements $\psi_k = \frac{1}{l_k} \sum_{j=1}^{s} C_{kj} q_j$. C_{kj} is an algebraic matrix of a projection of nodal displacements for the directions perpendicular to bar axes [34]. The above formalism leads to the linear system of algebraic equations $\sum_{j=1}^{s} (k_{ij} + k_{ij}^{G}) q_j = Q_i$, in which the linear stiffness matrix k_{ij} and geometric stiffness matrix k_{ij}^{G} can be expressed in algebraic form $k_{ij} = \sum_{k=1}^{e} B_{ki} \frac{E_k A_k}{l_k} B_{kj}$, $k_{ij}^{G} = \sum_{k=1}^{e} C_{ki} \frac{S_k}{l_k} C_{kj}$ (see references [33,34] for further details). The approach is not dependent on any approximation typical for the finite element method.

In a discrete model (DM), the strain energy of a tensegrity truss can be expressed in the matrix notation as a quadratic form of nodal displacements $\mathbf{q}$:

$$E_{\mathrm{s}}^{\mathrm{DM}} = \frac{1}{2} \mathbf{q}^{\mathrm{T}} \mathbf{K} \mathbf{q}, \tag{1}$$

where: $\mathbf{K} = \mathbf{K}_{\mathrm{L}} + \mathbf{K}_{\mathrm{G}}$, $\mathbf{K}_{\mathrm{L}}$-global linear stiffness matrix, $\mathbf{K}_{\mathrm{G}}$-global geometric stiffness matrix.

The self-equilibrated system of axial forces of the structure is represented by the geometric stiffness matrix and is related to the equations of equilibrium in the actual configuration of the lattice.

The strain energy of a solid according to the symmetric geometrically linear 3D elasticity theory (ET) [35] can be expressed as:

$$E_{\mathrm{s}}^{\mathrm{ET}} = \frac{1}{2} \int_{V} \boldsymbol{\varepsilon}^{\mathrm{T}} \mathbf{E} \boldsymbol{\varepsilon} \, dV, \tag{2}$$

where: $\boldsymbol{\varepsilon}$-vector of strain components, $\mathbf{E}$-elasticity matrix.

In order to analyze mechanical properties of the material, it is proposed to compare the strain energy of an unsupported tensegrity to the strain energy of a cube, with an assumption that the strain energy of the cube is constant in its volume. In a general case both the analyzed structure and the solid can have arbitrary dimensions. However, in order to show how the continuum model is constructed, a typical tensegrity module inscribed into a cube of edge length a is considered (presented in Figure 1 of reference [28]).

With the above assumptions, the strain energy of the cube of edge length a, according to the symmetric 3D elasticity theory (ET):

$$E_s^{\mathrm{ET}} = \frac{1}{2}\int_V \varepsilon^{\mathrm{T}}\mathbf{E}\varepsilon\, dV = \frac{1}{2}\varepsilon^{\mathrm{T}}\mathbf{E}\varepsilon\, a^3. \tag{3}$$

To compare the energies and build the equivalent elasticity matrix, the nodal displacements of the structure are expressed by the average mid-values of displacements and their derivatives in the center of the cube of edge length a, with the use of Taylor series expansion. Nodal coordinates of the analyzed tensegrity structure can be expressed using the parameter a, which corresponds to the edge length of the cube: $\{\alpha_{xi}a, \alpha_{yi}a, \alpha_{zi}a\}$. Then, the parameters of the node i (for example nodal displacements) can be described as:$\Delta x_i = \alpha_{xi}a$, $\Delta y_i = \alpha_{yi}a$, $\Delta z_i = \alpha_{zi}a$.

The next step of the analysis is a substitution of the determined nodal displacements in the formula (1). As a result, an expression containing a constant part (independent of a) and terms with a factor a^n, $n \in \{1, 2, \ldots\}$ is obtained. However, in the case of small values of a, terms with the factor a^n can be regarded as higher order terms in the Taylor series expansion and should be omitted. Small values of the module dimension a should be considered as relative to the total dimensions of the metamaterial. Moreover, the mentioned terms contain displacement derivatives greater than the first one, which is beyond the scope of the symmetric theory of elasticity.

Comparison of strain energies (1) and (3) leads to the determination of coefficients of the matrix **E**. In a general case of an anisotropic structure, the obtained elasticity matrix in Voight's notation [32] has the following form:

$$\mathbf{E} = \begin{bmatrix} e_{11} & e_{12} & e_{13} & e_{14} & e_{15} & e_{16} \\ & e_{22} & e_{23} & e_{24} & e_{25} & e_{26} \\ & & e_{33} & e_{34} & e_{35} & e_{36} \\ & & & e_{44} & e_{45} & e_{46} \\ & & & & e_{55} & e_{56} \\ sym. & & & & & e_{66} \end{bmatrix}. \tag{4}$$

It contains 36 coefficients, including 21 independent ones. The above matrix can take different and particular forms, with the type depending on eight possible symmetries [36]. The proposed continuum model is non-linear in the sense of equations of equilibrium considered in actual configuration. Numerical simulation and validation of the proposed continuum model is presented in the Appendix A at the end of the present paper. Other possible continuum models of the lattices are discussed in references [37,38].

3. Simplex-Based Tensegrity Lattices

The metamaterial considered in this paper is constructed from one of the most popular tensegrity modules—a 4-strut simplex (see references [7,9,28] for geometrical details). It is a typical tensegrity, which consists of four separate struts surrounded by the continuous system of twelve cables [7,8]. The 4-strut simplex module is obtained from a regular prism by rotating one of its bases 135 degrees clockwise or counter clockwise.

One of the special characteristics of tensegrity structures are infinitesimal mechanisms that are balanced with a self-equilibrated system of normal forces [7,8,10]. The considered simplex module has one infinitesimal mechanism and one corresponding self-stress state—self-stress is expressed by the relative forces in struts and cables with a multiplier S_0 (see reference [28] for the detailed description).

It should be noticed that the proposed unit cell is an anisotropic structure. However, it is proved below that the metamaterial based on such unit cells exhibits orthotropic properties.

Simplex tensegrity modules described above can be arranged in different patterns to form a material with certain properties. Depending on the type of the module used (with the basis rotated clockwise or counter clockwise) and the way in which the modules are connected, a material with different mechanical characteristics can be obtained. In the present paper a material with orthotropic properties is proposed, as it exhibits some special features (negative Poisson's ratio as an example [28]).

A system that consists of four simplex modules joined together by common cables of the lower bases and common nodes of the upper bases is presented in Figure 1a. The modules are arranged alternately: a module that is rotated clockwise is put next to the module with the counter-clockwise rotation. Although a single simplex module is anisotropic, the whole structure has orthotropic properties. Following this method, a regular eight-module supercell (Figure 1b [28]), constructed using two four-module layers, was considered. The upper layer of the supercell was built from the four-module layer turned upside-down and connected with the bottom layer using common cables.

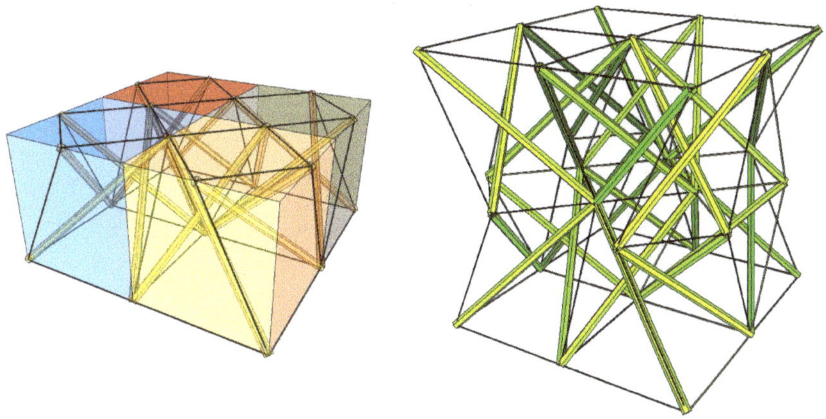

Figure 1. Geometry of a regular supercell (R): (**a**) A four-module supercell; (**b**) An eight module supercell [28].

The elasticity matrix $\mathbf{E}_R$ obtained from the continuum model of the considered regular supercell has the following form:

$$\mathbf{E}_R = \begin{bmatrix} e_{R11} & e_{R12} & e_{R13} & 0 & 0 & 0 \\ & e_{R11} & e_{R13} & 0 & 0 & 0 \\ & & e_{R33} & 0 & 0 & 0 \\ & & & e_{R12} & 0 & 0 \\ & & & & e_{R13} & 0 \\ sym. & & & & & e_{R13} \end{bmatrix}, \tag{5}$$

with the coefficients [28]:

$$\begin{aligned} e_{R11} &= \frac{2EA}{a^2}(0.314815 + 0.960318 \cdot k - 0.0794978 \cdot \sigma), \\ e_{R12} &= \frac{EA}{a^2}(0.2962963 + 0.353553 \cdot k - 0.0134742 \cdot \sigma), \\ e_{R13} &= \frac{EA}{a^2}(0.740741 + 0.268328 \cdot k + 0.17247 \cdot \sigma), \\ e_{R33} &= \frac{2EA}{a^2}(0.592593 + 1.07331 \cdot k - 0.17247 \cdot \sigma), \end{aligned}$$

where: $k = \frac{(EA)_{cable}}{(EA)_{strut}}$, $(EA)_{strut} = EA, \sigma = \frac{S}{EA}$.

The volume of the regular lattice is $V_R = aA_{strut}(48.00 + 62.15k)$.

Two other tensegrity lattices inscribed into a cube $2a \times 2a \times 2a$ are considered: with a small and large height-to-base area ratio of the module (Figures 2 and 3).

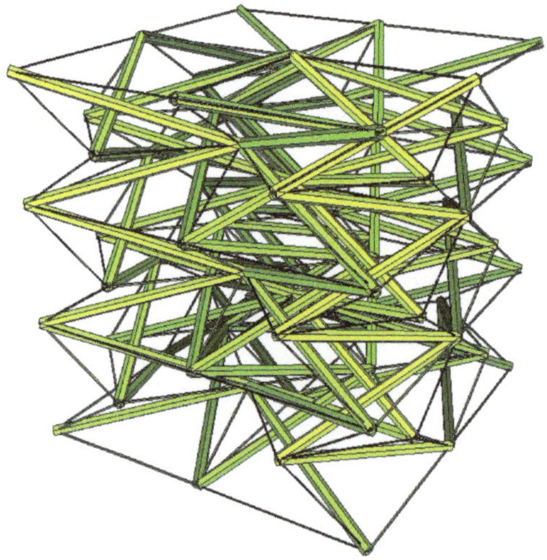

Figure 2. Tensegrity lattice with low modules (L).

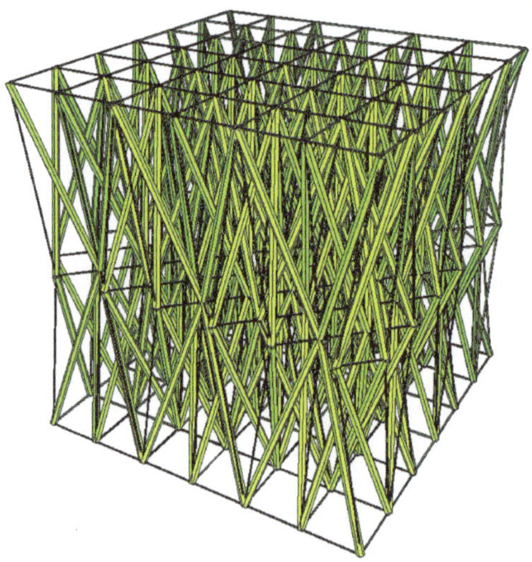

Figure 3. Tensegrity lattice with high modules (H).

Self-equilibrated forces for (L) and (H) models differ from the ones presented in Figure 1. The elasticity matrices $\mathbf{E}_L$ and $\mathbf{E}_H$ have orthotropic structures as before, but with different coefficients:

$$
\begin{aligned}
e_{L11} &= \frac{2EA}{a^2}(2.00729 + 2.67837 \cdot k - 0.135911 \cdot \sigma), \\
e_{L12} &= \frac{EA}{a^2}(1.88921 + 1.06066 \cdot k + 0.223898 \cdot \sigma), \\
e_{L13} &= \frac{EA}{a^2}(0.524781 + 0.576035 \cdot k + 0.0479246 \cdot \sigma), \\
e_{L33} &= \frac{2EA}{a^2}(0.0466472 + 0.256015 \cdot k - 0.0479246 \cdot \sigma),
\end{aligned}
$$

with the volume $V_L = aA_{strut}(112.00 + 125.21k)$ and

$$e_{H11} = \frac{2EA}{a^2}(0.0971325 + 2.28422 \cdot k - 0.217699 \cdot \sigma),$$
$$e_{H12} = \frac{EA}{a^2}(0.0914188 + 1.06066 \cdot k - 0.134146 \cdot \sigma),$$
$$e_{H13} = \frac{EA}{a^2}(2.05692 + 0.279923 \cdot k + 0.569545 \cdot \sigma),$$
$$e_{H33} = \frac{2EA}{a^2}(14.8098 + 10.0772 \cdot k - 0.569545 \cdot \sigma),$$

with the volume $V_H = aA_{strut} = 307.35 + 260.26k$.

The supercells are representative because the mechanical properties described in elasticity matrices do not change in repetitive bigger volumes.

The proposed tensegrity metameterials (L), (R) and (H) will be compared and analyzed in the next chapter according to the eigenvalues and eigenvectors of the elasticity matrices under the assumption of an equal total volume.

4. Extreme Properties of Tensegrity Lattices

The eigenvalues $\lambda_i > 0$ ($i = 1, 2, \ldots, 6$), and orthogonal eigenvectors $\mathbf{w}_i$ of elasticity matrices of the proposed metamaterials depend on the parameters k and σ. The eigenvalues in this chapter are calculated for the (L) and (H) elasticity matrices calibrated to an even volume of the material in each lattice (as in the metamaterial (R)).

Let us start the analysis from the lowest eigenvalues. The elasticity matrices are positive definite so only positive eigenvalues are to be considered. The values of $\lambda_{\min}$ for $k \in (0,1)$ and $\sigma \in (0,1)$ for low (L), regular (R) and high (H) metamaterials are presented in Figure 4.

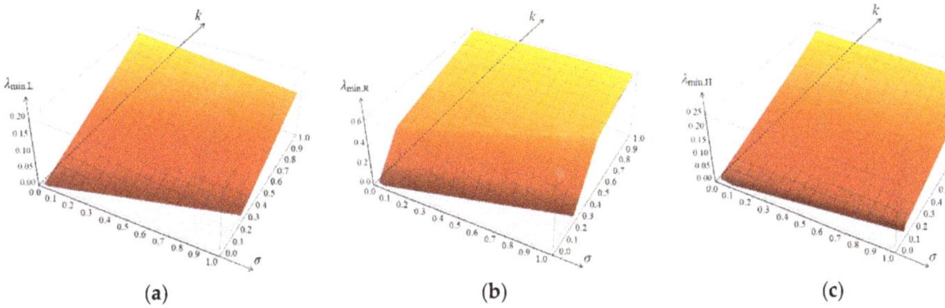

Figure 4. Distributions of lowest eigenvalue for: (a) Low; (b) Regular; (c) High metamaterial.

In the (L) metamaterial (Figure 4a) the lowest eigenvalue does not change the mode of deformation within the considered area. The metamaterial is unimode according to the classification of reference [29]. In the (R) metamaterial (Figure 4b) the change of the mode of deformation of the lowest eigenvalue is observed but far from the zero line. The material is also unimode for selected parameters. In the (H) metamaterial (Figure 4c) the change of the mode of deformation of the lowest eigenvalue is observed close to the zero line. It means that the material is close to bimode [29]. The lines of zero values representative for unimode soft deformation are presented in Figure 5 for all metamaterials together.

The highest eigenvalues of the matrices $\mathbf{E}$ depend on the parameters k and σ to a small extent (a couple of percent), therefore the analysis is not considered in this paper.

The second part of the analysis is proposed for the parameter $k = 0.1$ which is typical for standard cable-strut structures. The values of self-stress parameters necessary to obtain the zero lowest eigenvalue are then the following: $\sigma_L = 0.35856$, $\sigma_M = 0.37317$ and $\sigma_H = 0.76479$. The sets of six eigenvalues are presented in Figure 6.

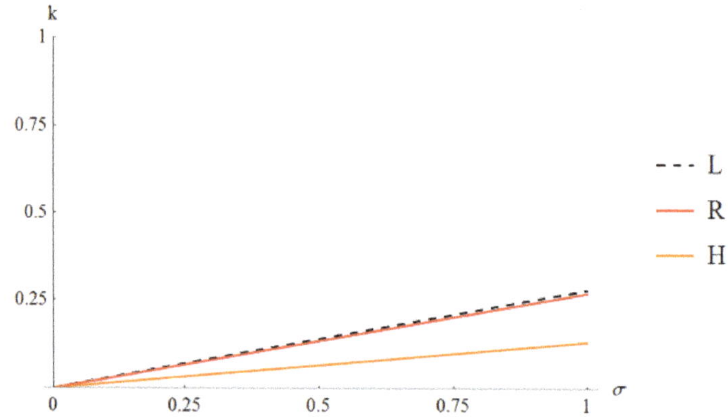

Figure 5. Lines of zero value of the lowest eigenvalue.

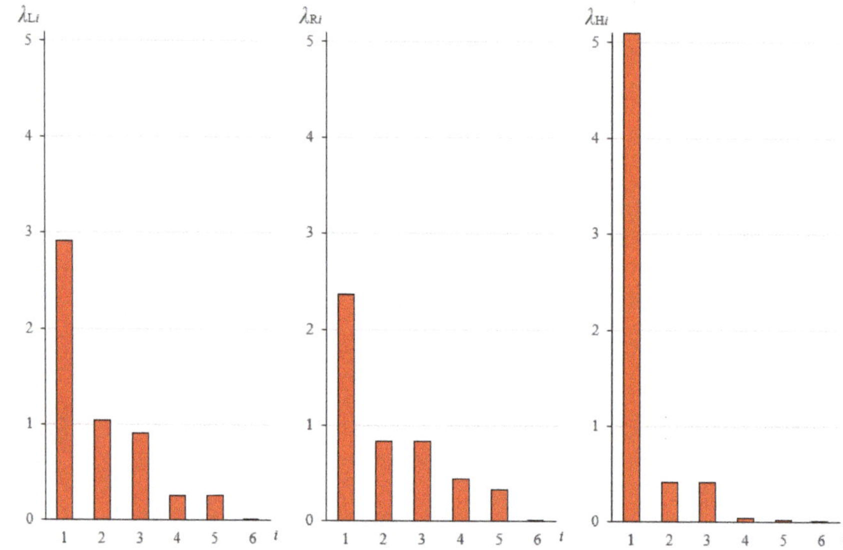

Figure 6. Distribution of eigenvalues for (L), (R) and (H) metamaterials.

As it can be noticed, the most extreme metamaterial is (H). One eigenvalue is high, three are relatively small and two are close to zero. The (H) metamaterial is nearly bimode. The materials (L) and (R) can be defined as unimode. The sequence of eigenvectors in the (L) metamaterial is different than in the others—the double eigenvalue which is responsible for shear vertical deformation is next to the zero eigenvalue in the material (L) and next to the highest eigenvalue in the materials (R) and (H). Six deformation modes can be defined in the metamaterials: stiff (extensional), soft (extensional), named "easy" in [29], medium extensional, high shear (double) and low shear. The results are presented graphically in Tables 1–5.

The analysis carried out above for the parameter $k = 0.1$ differs for other k values. A similar analysis can be performed on the basis of the matrix inverse to **E**, defining others that can be called stress modes instead of deformation modes.

Table 1. Stiff (extensional) mode of deformation.

$\lambda_{1L} = 2.90631$	$\lambda_{1R} = 2.36011$	$\lambda_{1H} = 5.09819$
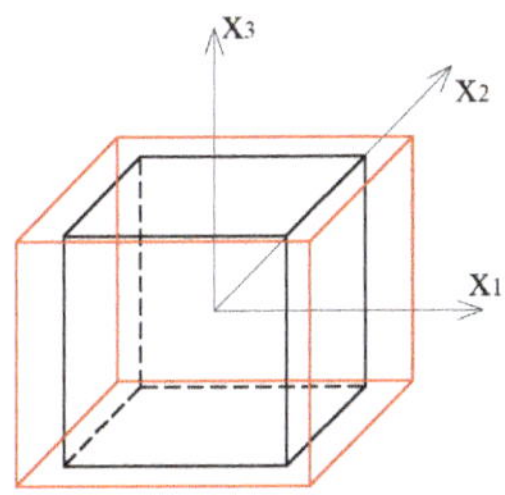	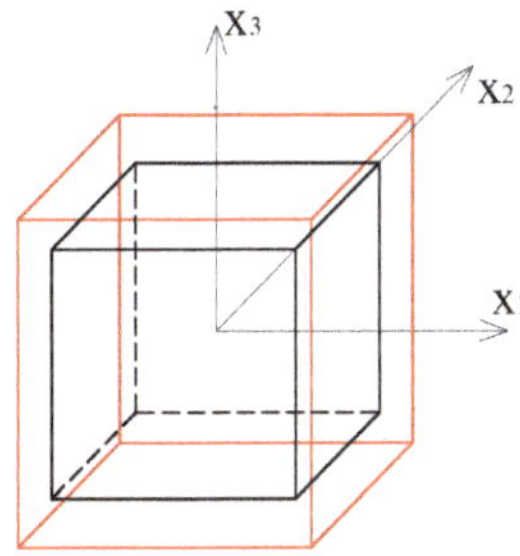	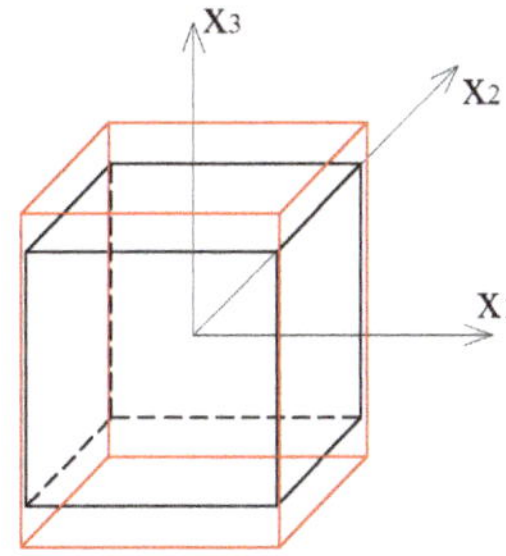
$\mathbf{w}_{1L} = \begin{bmatrix} 5.444189 \\ 5.444189 \\ 1 \\ 0 \\ 0 \\ 0 \end{bmatrix}$	$\mathbf{w}_{1R} = \begin{bmatrix} 0.654487 \\ 0.654487 \\ 1 \\ 0 \\ 0 \\ 0 \end{bmatrix}$	$\mathbf{w}_{1H} = \begin{bmatrix} 0.0819302 \\ 0.0819302 \\ 1 \\ 0 \\ 0 \\ 0 \end{bmatrix}$

Table 2. Soft (extensional) mode of deformation.

$\lambda_{6L} \sim 0$	$\lambda_{6R} \sim 0$	$\lambda_{6H} \sim 0$
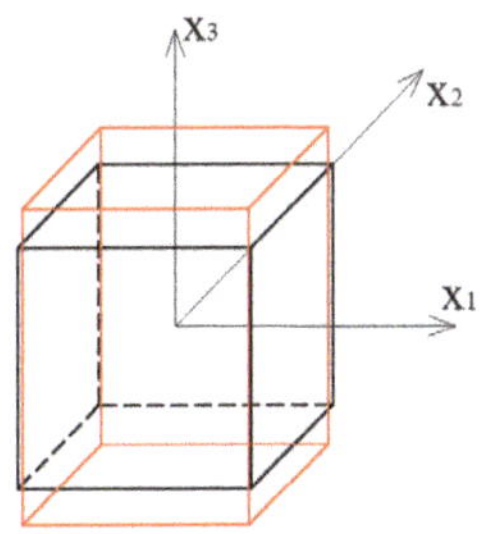	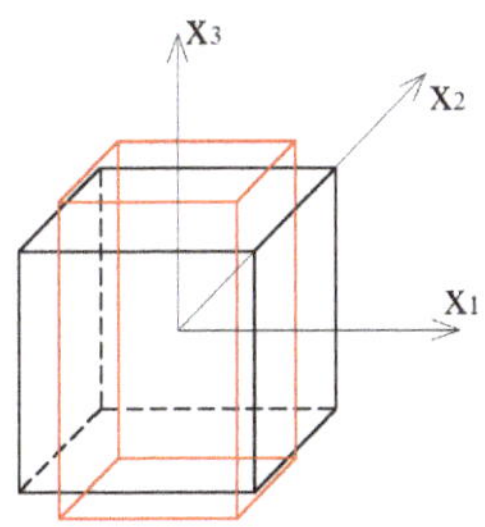	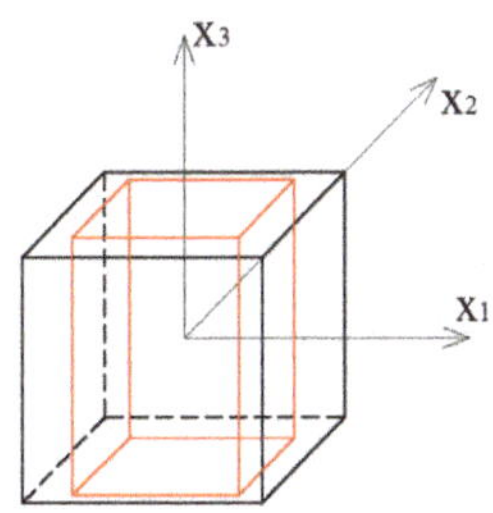
$\mathbf{w}_{6L} = \begin{bmatrix} -0.0918408 \\ -0.0918408 \\ 1 \\ 0 \\ 0 \\ 0 \end{bmatrix}$	$\mathbf{w}_{6R} = \begin{bmatrix} -0.763957 \\ -0.763957 \\ 1 \\ 0 \\ 0 \\ 0 \end{bmatrix}$	$\mathbf{w}_{6H} = \begin{bmatrix} -1 \\ -1 \\ 0.16386 \\ 0 \\ 0 \\ 0 \end{bmatrix}$

Table 3. Medium extensional mode of deformation.

$\lambda_{2L} = 1.04074$	$\lambda_{4R} = 0.435737$	$\lambda_{4H} = 0.0365033$
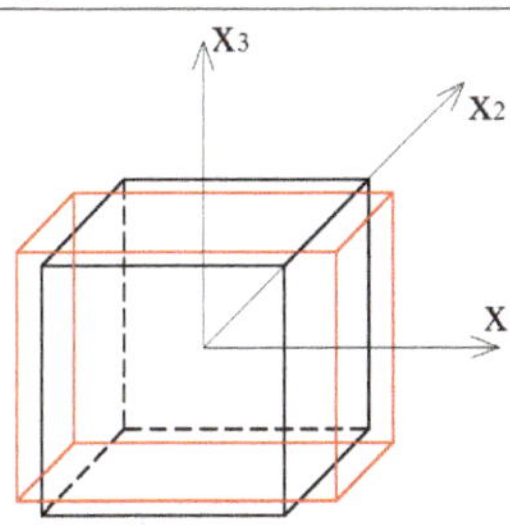	$\mathbf{w}_{2L} = \mathbf{w}_{4R} = \mathbf{w}_{4H} = \begin{bmatrix} 1 \\ -1 \\ 0 \\ 0 \\ 0 \\ 0 \end{bmatrix}$	

Table 4. High shear modes of deformation.

$\lambda_{4L} = \lambda_{5L} = 0.26249$	$\lambda_{2R} = \lambda_{3R} = 0.831934$	$\lambda_{2H} = \lambda_{3H} = 0.412162$
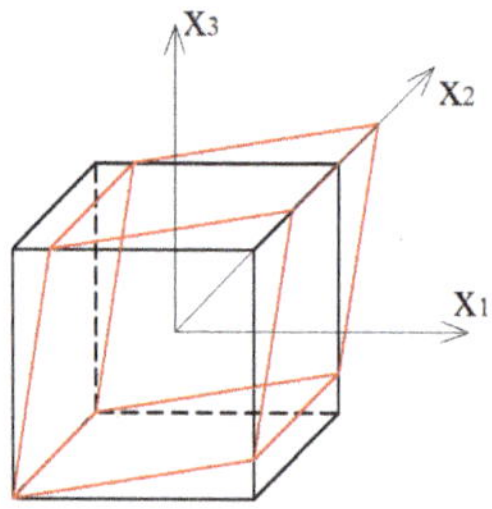	$\mathbf{w}_{4L} = \mathbf{w}_{2R} = \mathbf{w}_{2H} = \begin{bmatrix} 0 \\ 0 \\ 0 \\ 0 \\ 1 \\ 0 \end{bmatrix}$	
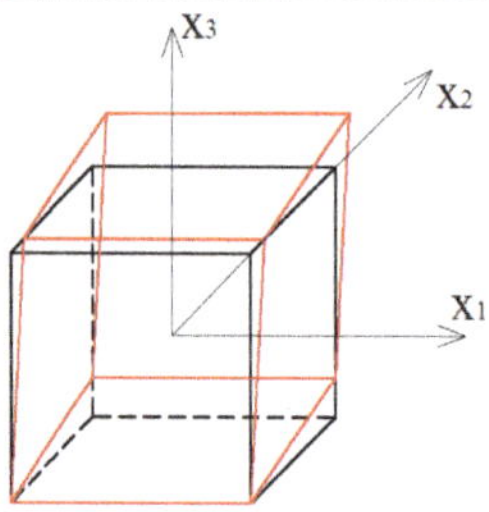	$\mathbf{w}_{5L} = \mathbf{w}_{3R} = \mathbf{w}_{3H} = \begin{bmatrix} 0 \\ 0 \\ 0 \\ 0 \\ 0 \\ 1 \end{bmatrix}$	

Table 5. Low shear mode of deformation.

$\lambda_{3L} = 0.908675$	$\lambda_{5R} = 0.326623$	$\lambda_{5H} = 0.015517$
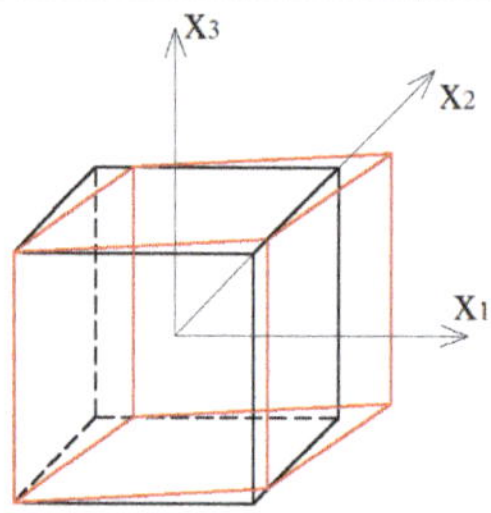	$\mathbf{w}_{5L} = \mathbf{w}_{5R} = \mathbf{w}_{5H} = \begin{bmatrix} 0 \\ 0 \\ 0 \\ 1 \\ 0 \\ 0 \end{bmatrix}$	

5. Conclusions

The present paper focuses on the analysis of a novel cellular metamaterial based on the simplex tensegrity pattern recognized as a metamaterial. Three tensegrity lattices are proposed, which differ in geometrical proportions. The authors use a continuum model of the lattices to estimate the influence of self-equilibrated normal forces and geometrical parameters on the behavior of the system. Moreover, a spectral analysis of the elasticity matrices, with the condition of the equal total volume of the lattices, is presented as a tool for comparing the metamaterials and defining their extreme properties.

The performed analyses allowed the authors to identify six typical deformation modes in the lattices, which were named stiff, soft and medium extensional modes, as well as high (double) and low shear modes. The proposed novel simplex based metamaterials can be regarded as extreme materials with unimode or nearly bimode mechanical properties. The occurrence of soft (easy) as well as stiff deformation modes allows for classifying the developed materials as smart tensegrity metamaterials, which is very promising as far as the engineering applications are concerned.

Author Contributions: Conceptualization, A.A.S.-Z. and W.G.; Methodology, A.A.S.-Z. and W.G.; Software, A.A.S.-Z.; Validation, A.A.S.-Z.; Formal Analysis, A.A.S.-Z. and W.G.; Investigation, A.A.S.-Z. and W.G.; Resources, W.G.; Data Curation, A.A.S.-Z.; Writing-Original Draft Preparation, A.A.S.-Z. and W.G.; Writing-Review & Editing, A.A.S.-Z. and W.G.; Visualization, A.A.S.-Z.; Supervision, W.G.

Funding: This research received no external funding.

Conflicts of Interest: The authors declare no conflict of interest.

Appendix A. Validation of the Continuum Model

In order to verify the proposed continuum model, its validation on an orthotropic structure was performed. The structure that was chosen for the verification of the model has a simple geometry and therefore, its behavior is easy to understand. The results obtained from the continuum model were compared with the ones from the discrete analysis. The calculations were made in the *Mathematica* software. The validation was performed on a space truss (Figure A1) shaped in such a way that it is an orthotropic structure. The orthotropy was obtained through diversification of the stiffness of particular truss members: members parallel to the axis x—stiffness EA, members parallel to the axis y—stiffness $2EA$, members parallel to the axis z—stiffness $3EA$, diagonal members—stiffness EA.

To validate the model, the following mechanical characteristics were compared: Young's moduli, Poisson's ratios and shear moduli. In the validation a geometric stiffness matrix representing the self-stress of the truss was considered. In order to simplify the calculations, so that the results could easily be compared, a certain level of self-stress $\sigma = S/EA = 0.5$ was assumed.

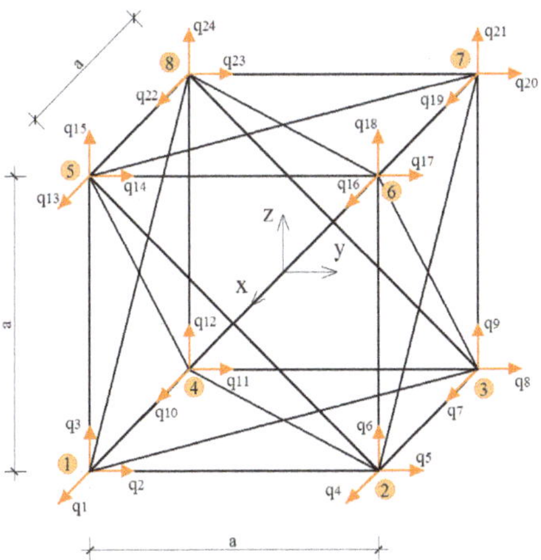

Figure A1. Space truss with 8 nodes and 24 degrees-of-freedom.

In the continuum analysis the structure was not supported, as it is not required in the proposed model. The following elasticity matrix **E** was obtained for the analyzed truss using the continuum model with self-stress included :

$$\mathbf{E} = \frac{EA}{a^2}\begin{bmatrix} 2\times\left(2+\sqrt{2}+2\sigma\right) & \sqrt{2}-2\sigma & \sqrt{2}-2\sigma & 0 & 0 & 0 \\ & 2\times\left(4+\sqrt{2}+2\sigma\right) & \sqrt{2}-2\sigma & 0 & 0 & 0 \\ & & 2\times\left(6+\sqrt{2}+2\sigma\right) & 0 & 0 & 0 \\ & & & \sqrt{2}+14\sigma & 0 & 0 \\ & & & & \sqrt{2}+14\sigma & 0 \\ sym. & & & & & \sqrt{2}+14\sigma \end{bmatrix}. \quad \text{(A1)}$$

The inverse matrix $\mathbf{H} = \mathbf{E}^{-1}$, with the assumption σ = 0.5 is the following:

$$\mathbf{H} = \frac{a^2}{EA}\begin{bmatrix} 0.113565 & -0.00357947 & -0.00270719 & 0 & 0 & 0 \\ -0.00357947 & 0.0781267 & -0.0018349 & 0 & 0 & 0 \\ -0.00270719 & -0.0018349 & 0.0595351 & 0 & 0 & 0 \\ 0 & 0 & 0 & 0.118847 & 0 & 0 \\ 0 & 0 & 0 & 0 & 0.118847 & 0 \\ 0 & 0 & 0 & 0 & 0 & 0.118847 \end{bmatrix}. \tag{A2}$$

Mechanical characteristics were calculated from the matrix **H** (A2), using the relations described in terms of mechanical coefficients of orthotropy [28,35]. For example, the analysis of a state σ_y = *const* leads to the determination of: Young's modulus E_2 and Poisson's ratios ν_{21} and ν_{23} in the following way:

$$\frac{1}{E_2} = 0.0781267\frac{a^2}{EA} \quad \Rightarrow \quad E_2 = 12.7997\,\frac{EA}{a^2}, \tag{A3}$$

$$-\frac{\nu_{21}}{E_2} = -0.00357947\frac{a^2}{EA} \quad \Rightarrow \quad \nu_{21} = 0.04582, \tag{A4}$$

$$-\frac{\nu_{23}}{E_2} = -0.0018349\frac{a^2}{EA} \quad \Rightarrow \quad \nu_{23} = 0.02349, \tag{A5}$$

In the same way, all other mechanical characteristics can be determined by analysing different stress states. The following values of mechanical properties were obtained:

$$\begin{array}{lll} & & \nu_{12} = 0.03152, \\ E_1 = 8.8055\,\frac{EA}{a^2}, & G_1 = 8.41421\,\frac{EA}{a^2}, & \nu_{13} = 0.02384, \\ E_2 = 12.7997\,\frac{EA}{a^2}, & G_2 = 8.41421\,\frac{EA}{a^2}, & \nu_{21} = 0.04582, \\ E_3 = 16.7968\,\frac{EA}{a^2}, & G_3 = 8.41421\,\frac{EA}{a^2}, & \nu_{23} = 0.02349, \\ & & \nu_{31} = 0.04547, \\ & & \nu_{32} = 0.03082. \end{array} \tag{A6}$$

Six typical stress states were analyzed using the discrete model. In each state the structure was supported in nodes in such a way that proper deformations could occur. The applied load caused in each case the stress of a unit value. The concentrated forces $P = \frac{1 \cdot a^2}{4}$ were applied to the proper nodes of the truss. For all analyzed stress states nodal displacements of the structure were determined and mechanical characteristics were calculated. In the analysis self-stress of the truss on the level of σ = 0.5 was applied.

Figure A2 shows the deformed truss in one of the analyzed states: σ_y = *const*, which corresponds to the uniform tension in the direction y. The adopted support scheme enables a free deformation of the structure in the considered uniform tension state.

The following values of nodal displacements were obtained in the analyzed state (the above values are divided by the factor a^3/EA):

$$\begin{array}{lll} q_1 = -0.00358, & q_2 = 0, & q_3 = 0, \\ q_4 = -0.00358, & q_5 = 0.07813, & q_6 = 0, \\ q_7 = 0, & q_8 = 0.07813, & q_9 = 0, \\ q_{10} = 0, & q_{11} = 0, & q_{12} = 0, \\ q_{13} = -0.00358, & q_{14} = 0, & q_{15} = -0.00183, \\ q_{16} = -0.00358, & q_{17} = 0.07813, & q_{18} = -0.00183, \\ q_{19} = 0, & q_{20} = 0.07813, & q_{21} = -0.00183, \\ q_{22} = 0, & q_{23} = 0, & q_{24} = -0.00183. \end{array} \tag{A7}$$

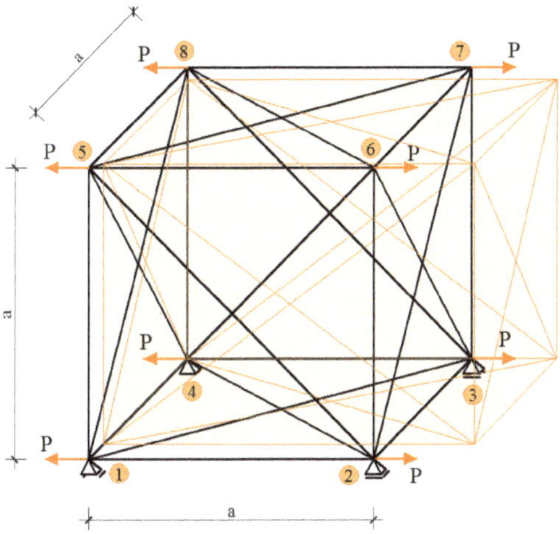

Figure A2. State σ_y = *const*—a deformed structure

The presented state σ_y = *const* was used to determine the values of: Young's modulus E_2 and Poisson's ratios ν_{21} and ν_{23} in the following way:

$$\begin{aligned} E_2 &= \frac{\sigma_y}{\varepsilon_y} = \frac{1 \cdot a}{q_5} = \frac{1 \cdot a}{0.07813} \cdot \frac{EA}{a^3} = 12.7997 \frac{EA}{a^2}, \\ \nu_{21} &= -\frac{\varepsilon_x}{\varepsilon_y} = -\frac{q_1}{q_5} = -\frac{-0.00358}{0.07813} = 0.04582, \\ \nu_{23} &= -\frac{\varepsilon_z}{\varepsilon_y} = -\frac{q_{15}}{q_5} = -\frac{-0.00183}{0.07813} = 0.02349. \end{aligned} \tag{A8}$$

It should be noticed that the above results are equal to the ones from the continuum model. Similar calculations were performed for other five stress states, in each case the results obtained from both analyses were fully consistent.

The validation of the proposed continuum model performed on a simple truss proved that it gives reliable results and therefore, it can be used to analyze extreme properties of various types of structures with self-stress.

References

1. Cui, T.J.; Smith, D.R.; Liu, R. *Metamaterials. Theory, Design and Applications*; Springer: London, UK, 2010.
2. Singh, G.; Ni, R.; Marwaha, A. A review of metamaterials and its applications. *Int. J. Eng. Trends Technol.* **2015**, *19*, 305–310. [CrossRef]
3. Akhras, G. Smart materials and smart systems for the future. *Canad. Milit. J.* **2000**, *1*, 25–32.
4. Al Sabouni-Zawadzka, A. Active control of smart tensegrity structures. *Arch. Civ. Eng.* **2014**, *4*, 517–534. [CrossRef]
5. Schenk, M.; Guest, S.D. Geometry of Miura-folded metamaterials. *Proc. Natl. Acad. Sci. USA* **2013**, *110*, 3276–3281. [CrossRef]
6. Lv, C.; Krishnaraju, D.; Konjevod, G.; Yu, H.; Jiand, H. Origami based mechanical metamaterials. *Sci. Rep.* **2014**, *4*, 5979. [CrossRef] [PubMed]
7. Skelton, R.E.; de Oliveira, M.C. *Tensegrity Systems*; Springer: New York, NY, USA; Dordrecht, The Netherlands; Heidelberg, Germany; London, UK, 2009.
8. Motro, R. *Tensegrity: Structural Systems for the Future*; Kogan Page Science: London, UK, 2003.
9. Wroldsen, A.S. Modelling and Control of Tensegrity Structures. Ph.D. Thesis, Department of Marine Technology, Norwegian University of Science and Technology, Trondheim, Norway, August 2007.

10. Calladine, C.R.; Pellegrino, S. First-order infinitesimal mechanisms. *Int. J. Solids Struct.* **1991**, *27*, 505–515. [CrossRef]
11. Gilewski, W.; Al Sabouni-Zawadzka, A. On possible applications of smart structures controlled by self-stress. *Arch. Civ. Mech. Eng.* **2015**, *15*, 469–478. [CrossRef]
12. Adam, B.; Smith, I.F.C. Self-diagnosis and self-repair of an active tensegrity structure. *J. Struct. Eng.* **2007**, *133*, 1752–1761. [CrossRef]
13. Al Sabouni-Zawadzka, A.; Gilewski, W. Inherent smartness of tensegrity structures—Structural elements applications. In Proceedings of the International Association for Shell and Spatial Structures (IASS), Amsterdam, The Netherlands, 17–20 August 2015.
14. Al Sabouni-Zawadzka, A.; Gilewski, W. Inherent smartness of tensegrity structures. *Appl. Sci.* **2018**, *8*, 787. [CrossRef]
15. Fraternali, F.; Carpentieri, G.; Amendola, A.; Skelton, R.E.; Nesterenko, V.F. Multiscale tunability of solitary wave dynamics in tensegrity metamaterials. *Appl. Phys. Lett.* **2014**, *105*, 201903. [CrossRef]
16. Fabbrocino, F.; Carpentieri, G.; Amendola, A.; Penna, R.; Fraternali, F. Accurate numerical methods for studying the nonlinear wave-dynamics of tensegrity metamaterials. *Eccomas Procedia Compdyn.* **2017**. [CrossRef]
17. Amendola, A.; Krushynska, A.; Daraio, C.; Pugno, N.M.; Fraternali, F. Tuning frequency band gaps of tensegrity mass-spring chains with local and global prestress. *Int. J. Solids Struct.* **2019**, in press. [CrossRef]
18. Wang, Y.T.; Liu, X.N.; Zhu, R.; Hu, G.K. Wave propagation in tunable lightweight tensegrity metastructure. *Sci. Rep.* **2018**, *8*, 11482. [CrossRef] [PubMed]
19. De Tommasi, D.; Marano, G.C.; Puglisi, G.; Trentadue, F. Optimal complexity and fractal limits of self-similar tensegrities. *Proc. R. Soc. A* **2016**, *471*, 20150250. [CrossRef]
20. De Tommasi, D.; Marano, G.C.; Puglisi, G.; Trentadue, F. Morphological optimization of tensegrity-type metamaterials. *Compos. Part B* **2017**, *115*, 182–187. [CrossRef]
21. Modano, M.; Mascolo, I.; Fraternali, F.; Bieniek, Z. Numerical and analytical approaches to the self-equilibrium problem of class teta = 1 tensegrity metamaterials. *Front. Mech.* **2018**, *5*, 5.
22. Fraternali, F.; Carpentieri, G.; Amendola, A. On the mechanical modeling of the extreme softening/stiffening response of axially loaded tensegrity prisms. *J. Mech. Phys. Solids* **2015**, *74*, 136–157. [CrossRef]
23. Rimoli, J.J.; Pal, R.K. Mechanical response of 3-dimensional tensegrity lattices. *Compos. Part B* **2017**, *115*, 30–42. [CrossRef]
24. Salahsoor, H.; Pal, R.K.; Rimoli, J.J. Material symmetry phase transitions in the three-dimensional tensegrity metamaterial. *J. Mech. Phys. Solids* **2018**, *119*, 382–399. [CrossRef]
25. Zhang, L.Y.; Li, S.X.; Zhu, S.X.; Zhang, B.Y.; Xu, G.K. Automatically assembled large-scale tensegrities by truncated regular polyhedral and prismatic elementary cells. *Compos. Struct.* **2018**, *184*, 30–40. [CrossRef]
26. Zhang, Q.; Zhang, D.; Dobah, Y.; Scarpa, F.; Fraternali, F.; Skelton, R.E. Tensegrity cell mechanical metamaterial with metal rubber. *Appl. Phys. Lett.* **2018**, *113*, 031906. [CrossRef]
27. Ma, Y.; Zhang, Q.; Dobah, J.; Scarpa, F.; Fraternali, F.; Skelton, R.E.; Zhang, D.; Hong, J. Meta-tensegrity: Design of a tensegrity prism with metal rubber. *Compos. Struct.* **2018**, *206*, 644–657. [CrossRef]
28. Al Sabouni-Zawadzka, A.; Gilewski, W. Smart metamaterial based on the simplex tensegrity pattern. *Materials* **2018**, *11*, 673. [CrossRef]
29. Milton, G.; Cherkaev, A.V. Which elasticity tensors are realizable? *J. Eng. Mater. Technol.* **1995**, *117*, 483–493. [CrossRef]
30. Kadic, M.; Buckmann, T.; Stenger, N.; Thiel, M. On the practicability of pentamode mechanical metamaterials. *Appl. Phys. Lett.* **2012**, *100*, 191901. [CrossRef]
31. Kasprzak, A.; Gilewski, W. 3D continuum model of tensegrity modules with the effect of self-stress. In Proceedings of the 11th World Congress on Computational Mechanics (WCCM XI, ECCM V), Barcelona, Spain, 20–25 July 2014.
32. Pellegrino, S.; Calladine, C.R. Matrix analysis of statically and kinematically indeterminate frameworks. *Int. J. Solids Struct.* **1990**, *22*, 409–422. [CrossRef]
33. Lewiński, T. On algebraic equations of elastic trusses, frames and grillages. *J. Theor. Appl. Mech.* **2001**, *39*, 307–322.

34. Pełczński, J.; Gilewski, W. An extension of algebraic equations of elastic trusses with self-equilibrated system of forces. In Proceedings of the 7th European Conference on Computational Fluid Dynamics, Glasgow, UK, 11–15 June 2018.
35. Green, A.E.; Zerna, W. *Theoretical Elasticity*; Oxford University Press: Oxford, UK, 1968.
36. Chadwick, P.; Vianello, M.; Cowin, S. A new proof that the number of linear elastic symmetries is eight. *J. Mech. Phys. Solids* **2001**, *49*, 2471–2492. [CrossRef]
37. Pal, R.K.; Ruzzene, M.; Rimoli, J. A continuum model for nonlinear lattices under large deformations. *Int. J. Solids Struct.* **2016**, *96*, 300–319. [CrossRef]
38. Pal, R.K.; Rimoli, J.; Ruzzene, M. Effect of large deformation pre-loads on the wave properties of hexagonal lattice. *Smart Mater. Struct.* **2016**, *25*, 054010. [CrossRef]

MDPI
St. Alban-Anlage 66
4052 Basel
Switzerland
Tel. +41 61 683 77 34
Fax +41 61 302 89 18
www.mdpi.com

Materials Editorial Office
E-mail: materials@mdpi.com
www.mdpi.com/journal/materials

CPSIA information can be obtained
at www.ICGtesting.com
Printed in the USA
LVHW070619280323
742728LV00006B/699

9 783039 360147